An Introduction to Structural Optimization

SOLID MECHANICS AND ITS APPLICATIONS
Volume 153

Series Editor: G.M.L. GLADWELL
Department of Civil Engineering
University of Waterloo
Waterloo, Ontario, Canada N2L 3GI

Aims and Scope of the Series

The fundamental questions arising in mechanics are: *Why?, How?,* and *How much?* The aim of this series is to provide lucid accounts written by authoritative researchers giving vision and insight in answering these questions on the subject of mechanics as it relates to solids.

The scope of the series covers the entire spectrum of solid mechanics. Thus it includes the foundation of mechanics; variational formulations; computational mechanics; statics, kinematics and dynamics of rigid and elastic bodies: vibrations of solids and structures; dynamical systems and chaos; the theories of elasticity, plasticity and viscoelasticity; composite materials; rods, beams, shells and membranes; structural control and stability; soils, rocks and geomechanics; fracture; tribology; experimental mechanics; biomechanics and machine design.

The median level of presentation is the first year graduate student. Some texts are monographs defining the current state of the field; others are accessible to final year undergraduates; but essentially the emphasis is on readability and clarity.

For other titles published in this series, go to
www.springer.com/series/6557

Peter W. Christensen · Anders Klarbring

An Introduction to Structural Optimization

Springer

Peter W. Christensen, Anders Klarbring
Division of Mechanics
Linköping University
SE-581 83 Linköping
Sweden
peter.christensen@liu.se, anders.klarbring@liu.se

ISBN 978-90-481-7947-3 e-ISBN 978-1-4020-8666-3

Printed on acid-free paper.

9 8 7 6 5 4 3 2 1

springer.com

Preface

This book has grown out of lectures and courses given at Linköping University, Sweden, over a period of 15 years. It gives an introductory treatment of problems and methods of structural optimization. The three basic classes of geometrical optimization problems of mechanical structures, i.e., size, shape and topology optimization, are treated. The focus is on concrete numerical solution methods for discrete and (finite element) discretized linear elastic structures. The style is explicit and practical: mathematical proofs are provided when arguments can be kept elementary but are otherwise only cited, while implementation details are frequently provided. Moreover, since the text has an emphasis on geometrical design problems, where the design is represented by continuously varying—frequently very many—variables, so-called first order methods are central to the treatment. These methods are based on sensitivity analysis, i.e., on establishing first order derivatives for objectives and constraints. The classical first order methods that we emphasize are CONLIN and MMA, which are based on explicit, convex and separable approximations. It should be remarked that the classical and frequently used so-called optimality criteria method is also of this kind. It may also be noted in this context that zero order methods such as response surface methods, surrogate models, neural networks, genetic algorithms, etc., essentially apply to different types of problems than the ones treated here and should be presented elsewhere. The numerical solutions that are presented are all obtained using in-house programs, some of which can be downloaded from the book's homepage at www.mechanics.iei.liu.se/edu_ug/strop/. These programs should also be used for solving some of the more extensive exercises provided.

The text is written for students with a background in solid and structural mechanics with a basic knowledge of the finite element method, although in our experience such knowledge could be replaced by a certain mathematical maturity. Previous exposure to basic optimization theory and convex programming is helpful but not strictly necessary.

The first three chapters of the book represent an introductory and preparatory part. In Chap. 1 we introduce the basic idea of mathematical design optimization and indicate its place in the broader frame of product realization, as well as define basic concepts and terminology. Chapter 2 is devoted to a series of small-scale problems that, on the one hand, give familiarity with the type of problems encountered in structural optimization and, on the other hand, are used as model problems in upcoming chapters. Chapter 3 reviews basic concepts of convex analysis, and exemplifies these by means of concepts from structural mechanics. Chapter 4 is, from an algorithmic point of view, the core chapter of the book. It introduces the basic idea of sequential explicit convex approximations, and CONLIN and MMA are presented. In Chap. 5 the latter is applied to stiffness optimization of a truss. This is a classical model problem of structural optimization which we investigate thoroughly. Chapter 6 concerns sensitivity analysis for finite element discretized structures. Sensitivities for shape changes are combined with two-dimensional shape representations such as Bézier and B-splines in Chap. 7, and this closes the treatment of shape optimization. Chapter 8 is essentially a preparation for the formulation of the problem of stiffness topology optimization. We review some classical results of the calculus of variations, and derive optimality conditions for stiffness optimization of distributed parameter systems. In Chap. 9 this problem is slightly extended and discretized, and it provides a gateway into the problem of topology optimization for continuous structures. We derive the optimality criteria method as a special case of the general explicit convex approximation method, discuss well-posedness and different types of regularization methods.

This being an introductory treatment, we have not made an effort to give a complete set of references, nor an historical account of structural optimization. For that we refer to existing monographs such as Haftka and Gürdal [18], Kirsch [22] and Bendsøe and Sigmund [4].

As mentioned, this book has its roots in several series of lectures at Linköping University, where the first one was given by the second author of this book in 1992. Following these, in the year 2000, a separate course in structural optimization was established, and Joakim Petersson was made responsible and defined its basic contents. After having taught the course on two occasions, Joakim very unexpectedly and sadly passed away in September 2002, [3]. The authors of this book then took over and shared responsibility for the course, initially teaching it in a way that was very close to the lecture notes of Joakim. Out of appreciation, we have continued to teach the course, and eventually written this book, closely following the spirit and style of Joakim, as we remember and understand it.

We like to extend a special thanks to Bo Torstenfelt and Thomas Borrvall for having provided some of the numerical solutions presented in the book. Torstenfelt's easy-to-use finite element program TRINITAS may be downloaded from the book's homepage, and should be used for two computer exercises on shape and topology optimization. A Java applet by Borrvall for performing topology optimization is also available on the homepage. For the permission to use their programs we are sincerely grateful.

Linköping,
July 2008

Peter W. Christensen
Anders Klarbring

Contents

Chapter 1
Introduction

This chapter introduces basic ideas and terminology of structural optimization. The role of mathematical design optimization in the product design process is discussed. Nested and simultaneous formulations of structural optimization, as well as the three basic geometric design parameterizations—size, shape and topology, are defined.

1.1 The Basic Idea

A *structure* in mechanics is defined by J.E. Gordon [17] as "any assemblage of materials which is intended to sustain loads." *Optimization* means making things the best. Thus, *structural optimization* is the subject of making an assemblage of materials sustain loads in the best way. To fix ideas, think of a situation where a load is to be transmitted from a region in space to a fixed support as in Fig. 1.1. We want to find the structure that performs this task in the best possible way. However, to make any sense out of that objective we need to specify the term "best." The first such specification that comes to mind may be to make the structure as light as possible, i.e., to minimize weight. Another idea of "best" could be to make the structure as stiff as possible, and yet another one could be to make it as insensitive to buckling or instability as possible. Clearly such maximizations or minimizations cannot be performed without any constraints. For instance, if there is no limitation on the amount of material that can be used, the structure can be made stiff without limit and we have an optimization problem without a well defined solution. Quantities that are usually constrained in structural optimization problems are stresses, displacements and/or the geometry. Note that most quantities that one can think of as constraints could also be used as measures of "best," i.e., as objective functions. Thus, one can put down a number of measures on structural performance—weight, stiffness, critical load, stress, displacement and geometry—and a structural optimization problem is formulated by picking one of these as an objective function that should be maximized or minimized and using some of the other measures as constraints. In Sect. 1.3 we will be specific about how such a formulation looks in mathematical terms. In the next section, Sect. 1.2, we will temporarily move the perspective in the other direction, and look at how structural optimization enters a broader picture.

1.2 The Design Process

The measures on structural performance indicated above are purely mechanical, e.g., we did not consider functionality, economy or esthetics. To make clear the

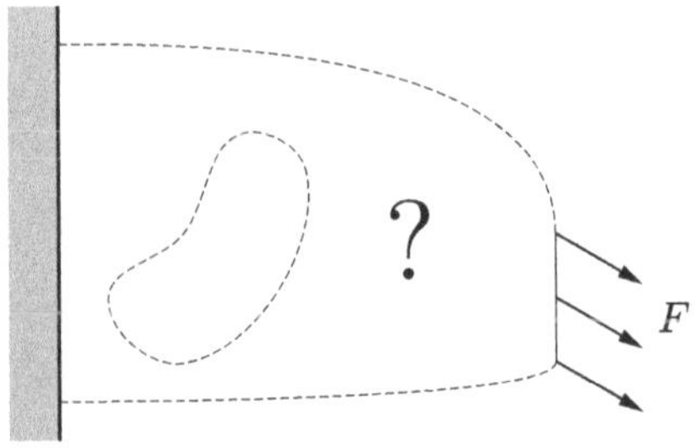

Fig. 1.1 Structural optimization problem. Find the structure which best transmits the load F to the support

position of structural optimization in relation to such, usually not mathematically defined, factors, we give a short indication of the main steps in the process of designing a product in general, as described by Kirsch [22]. At least in an ideal world these steps are as follows:

1. *Function*: What is the use of the product? Think of the design of a bridge: how long and broad should it be, how many driving lanes, what loads can be expected, etc.?
2. *Conceptual design*: What type of construction concept should we use? When we are to construct a bridge we need to decide if we are to build a truss bridge, a suspension bridge or perhaps an arch bridge.
3. *Optimization*: Within the chosen concept, and within the constraints on function, make the product as good as possible. For a bridge it would be natural to minimize cost; perhaps indirectly by using the least possible amount of material.
4. *Details*: This step is usually controlled by market, social or esthetic factors. In the bridge case, perhaps we need to choose an interesting color.

The traditional, and still dominant, way of realizing step 3 is the *iterative-intuitive* one, which can be described as follows. (a) A specific design is suggested. (b) Requirements based on the function are investigated. (c) If they are not satisfied, say the stress is too large, a new design must be suggested, and even if such requirements are satisfied the design may not be optimal (the bridge may be overly heavy) so we still may want to suggest a new design. (d) The suggested new design is brought back to step (b). In this way an iterative process is formed where, on mainly intuitive grounds, a series of designs are created which hopefully converges to an acceptable final design.

For mechanical structures, step (b) of the *iterative-intuitive* realization of step 3, is today almost exclusively performed by means of computer based methods like the Finite Element Method (FEM) or Multi Body Dynamics (MBD). These methods imply that every design iteration can be analyzed with greater confidence, and probably every step can be made more effective. However, they do not lead to a basic change of the strategy.

The *mathematical design optimization* method is conceptually different from the iterative-intuitive one. In this method a mathematical optimization problem is formulated, where requirements due to the function act as constraints and the concept "as good as possible" is given precise mathematical form. Thus, step 3 in the design process is much more automatic in mathematical design optimization than in an iterative-intuitive approach.

This text is concerned with a subset of the field of mathematical design optimization. That is, we treat mechanical structures whose main task is to carry loads. This subset is termed *structural optimization*.

Clearly, not all factors can be usefully treated in a mathematical design optimization method. A basic requirement is that the factor need to be measurable in mathematical form. This is usually easy for mechanical factors but difficult for, say, esthetic ones.

1.3 General Mathematical Form of a Structural Optimization Problem

The following function and variables are always present in a structural optimization problem:

- *Objective function* (f): A function used to classify designs. For every possible design, f returns a number which indicates the goodness of the design. Usually we choose f such that a small value is better than a large one (a minimization problem). Frequently f measures weight, displacement in a given direction, effective stress or even cost of production.
- *Design variable* (x): A function or vector that describes the design, and which can be changed during optimization. It may represent geometry or choice of material. When it describes geometry, it may relate to a sophisticated interpolation of shape or it may simply be the area of a bar, or the thickness of a sheet.
- *State variable* (y): For a given structure, i.e., for a given design x, y is a function or vector that represents the response of the structure. For a mechanical structure, response means displacement, stress, strain or force.

A general structural optimization problem now takes the form:

$$(\mathbb{SO}) \quad \begin{cases} \text{minimize } f(x, y) \text{ with respect to } x \text{ and } y \\ \text{subject to} \quad \begin{cases} \text{behavioral constraints on } y \\ \text{design constraints on } x \\ \text{equilibrium constraint.} \end{cases} \end{cases}$$

One can certainly imagine a problem with several objective functions, a so-called *multiple criteria*, or *vector* optimization problem:

$$\text{minimize } (f_1(x, y), f_2(x, y), \ldots, f_l(x, y)), \tag{1.1}$$

where l is the number of objective functions, and the constraints are the same as for ($\mathbb{SO}$). This is not a standard optimization problem since all f_i:s in general are not minimized for the same x and y. Instead, one therefore typically tries to achieve so-called *Pareto optimality*: a design is Pareto optimal if there does not exist any other design that satisfies all of the objectives better. Thus, (x^*, y^*) satisfying the constraints is Pareto optimal if there is no other (x, y) satisfying the constraints such that

$$f_i(x, y) \leq f_i(x^*, y^*), \quad \text{for all } i = 1, \ldots, l,$$
$$f_i(x, y) < f_i(x^*, y^*), \quad \text{for at least one } i \in \{1, \ldots, l\}.$$

The most common way to obtain a Pareto optimal point of (1.1) is to form a scalar objective function

$$\sum_{i=1}^{l} w_i f_i(x, y), \tag{1.2}$$

where $w_i \geq 0,\ i = 1, \ldots, l$, are so-called weight factors satisfying $\sum_{i=1}^{l} w_i = 1$. The problem of minimizing (1.2) under the constraints in $(\mathbb{SO})$ is a standard scalar optimization problem, the solution of which is a Pareto optimum of (1.1). By varying the weights, different Pareto optima are obtained. It should be remarked, however, that in general not every Pareto optimal point may be obtained with this simple method.

In this text we will consider only structural optimization problems of the form $(\mathbb{SO})$, i.e. problems with only one scalar objective function. The reader is referred to Ehrgott and Gandibleux [14], and the references therein, for a thorough discussion of multicriteria optimization.

Three types of constraints are indicated in $(\mathbb{SO})$: (1) *Behavioral constraints* are constraints on the state variable y. Usually they are written $g(y) \leq 0$, where g is a function which represents, e.g., a displacement in a certain direction. (2) *Design constraints* are similar constraints involving the design variable x. Obviously, these two types of constraints can be combined. Finally, in a naturally discrete problem or a discretized problem that is linear (we will discuss these two types of problems in Sect. 1.5), the *equilibrium constraint* looks like

$$\boldsymbol{K}(x)\boldsymbol{u} = \boldsymbol{F}(x), \tag{1.3}$$

where $\boldsymbol{K}(x)$ is the stiffness matrix of the structure, which generally is a function of the design, $\boldsymbol{u}$ is the displacement vector and $\boldsymbol{F}(x)$ is the force vector which may also depend on the design. Note that the displacement vector $\boldsymbol{u}$ takes the role of the general state variable y. In a continuum problem, the equilibrium constraint will typically be a partial differential equation. Moreover, in a dynamic structural optimization problem, equilibrium should be seen as dynamic equilibrium. A broader term than equilibrium constraint that encompasses this is *state problem.*

In the formulation $(\mathbb{SO})$, y and x are treated as independent variables. Such a formulation is usually called a *simultaneous formulation*, since equilibrium (or more generally, the state problem) is solved simultaneously with the optimization problem. However, a very frequent situation is that the state problem uniquely defines y in case of a given x, e.g., if $\boldsymbol{K}(x)$ is invertible for all x; we have $\boldsymbol{u} = \boldsymbol{u}(x) = \boldsymbol{K}(x)^{-1}\boldsymbol{F}(x)$. By treating $\boldsymbol{u}(x)$ as a given function, the equilibrium constraint can be left out of $(\mathbb{SO})$, and this function can be substituted for the state variable, which gives

$$(\mathbb{SO})_{\text{nf}} \quad \begin{cases} \min\limits_{x} & f(x, \boldsymbol{u}(x)) \\ \text{s.t.} & g(x, \boldsymbol{u}(x)) \leq 0, \end{cases}$$

where s.t. denotes "subject to," and we have assumed that all state and design constraints can be written as $g(x, \boldsymbol{u}) \leq 0$. This formulation is called a *nested formulation* and will be the starting point for numerical methods presented in this text.

When treating $(\mathbb{SO})_{\text{nf}}$ numerically, one usually needs derivatives of f and g with respect to the design x. To find such derivatives is the goal of *sensitivity analysis*. Since the function $\boldsymbol{u}(x)$ is only implicitly given, this is generally a nontrivial task.

1.4 Three Types of Structural Optimization Problems

In this text, x will almost exclusively represent some sort of geometric feature of the structure. Depending on the geometric feature, we divide structural optimization problems into three classes:

- *Sizing optimization*: This is when x is some type of structural thickness, i.e., cross-sectional areas of truss members, or the thickness distribution of a sheet. A sizing optimization problem for a truss structure is shown in Fig. 1.2.
- *Shape optimization*: In this case x represents the form or contour of some part of the boundary of the structural domain. Think of a solid body, the state of which is described by a set of partial differential equations. The optimization consists in choosing the integration domain for the differential equations in an optimal way. Note that the connectivity of the structure is not changed by shape optimization: new boundaries are not formed. A two-dimensional shape optimization problem is seen in Fig. 1.3.
- *Topology optimization*: This is the most general form of structural optimization. In a discrete case, such as for a truss, it is achieved by taking cross-sectional areas of truss members as design variables, and then allowing these variables to take

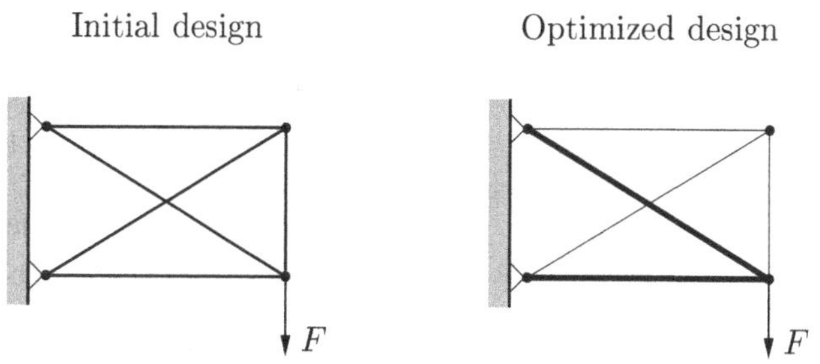

Fig. 1.2 A sizing structural optimization problem is formulated by optimizing the cross-sectional areas of truss members

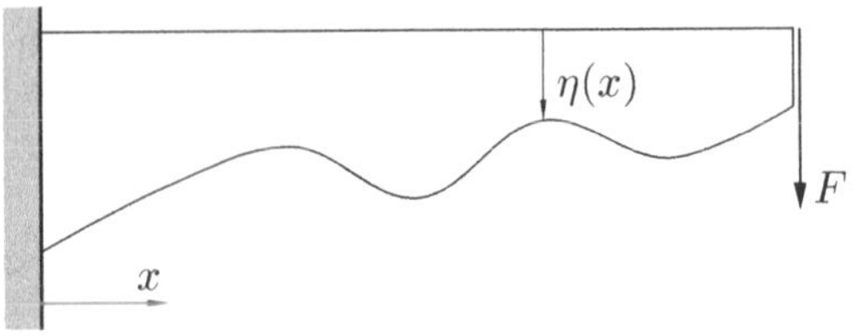

Fig. 1.3 A shape optimization problem. Find the function $\eta(x)$, describing the shape of the beam-like structure

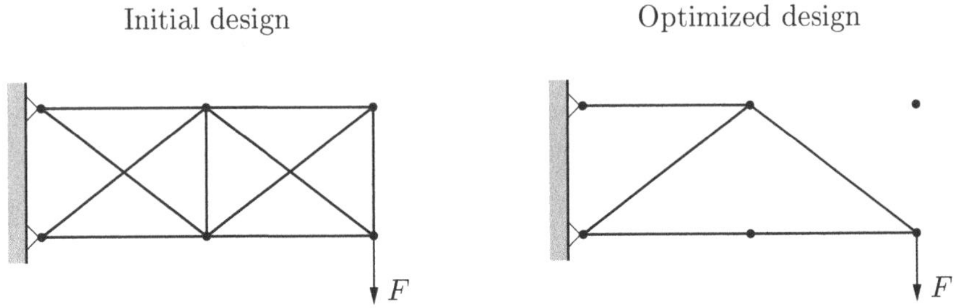

Fig. 1.4 Topology optimization of a truss. Bars are removed by letting cross-sectional areas take the value zero

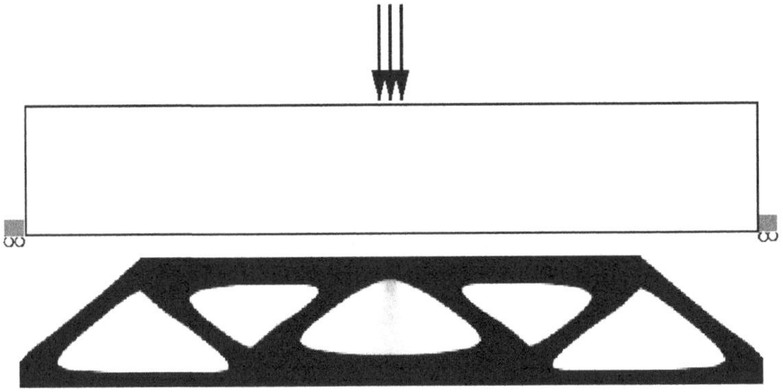

Fig. 1.5 Two-dimensional topology optimization. The box is to be filled to 50% by material. Where should the material be placed for optimal performance under loads and boundary conditions shown in the *upper picture*? The result is shown in the *second picture*. Calculations performed by Borrvall

the value zero, i.e., bars are removed from the truss. In this way the connectivity of nodes is variable so we may say that the topology of the truss changes, see Fig. 1.4. If instead of a discrete structure we think of a continuum-type structure such as a two-dimensional sheet, then topology changes can be achieved by letting the thickness of the sheet take the value zero. If pure topological features are optimized, the optimal thickness should take only two values: 0 and a fixed maximum sheet thickness. In a three-dimensional case the same effect can be achieved by letting x be a density-like variable that can only take the values 0 and 1. Figure 1.5 shows an example of topology optimization.

Ideally, shape optimization is a subclass of topology optimization, but practical implementations are based on very different techniques, so the two types are treated separately in this text and elsewhere. Concerning the relation between topology and sizing optimization, the situation is the opposite: from a fundamental point of view they are very different, but they are closely related from practical considerations.

When the state problem is a differential equation, we can say that shape optimization concerns control of the domain of the equation, while sizing and topology optimization concern control of its parameters.

The fact that there exist several types of structural optimization problems seems to have two different interpretations in terms of the design process of Sect. 1.2. The first one is that the boundary between step 2 and step 3 is flexible: topology optimization, which is the most general type of structural optimization, requires a less detailed description of the concept than, e.g., shape optimization. The other possible interpretation is that we have only partially left the intuitive-iterative method when doing structural optimization: an intuitive ingredient is left and it is likely that several different types of structural optimization problems need to be solved before step 3 is finished.

1.5 Discrete and Distributed Parameter Systems

As already indicated in previous sections, the design variable x and the state variable u may, depending on the situation, be finite dimensional (i.e., they belong to the space $\mathbb{R}^n$ of n-tuples of real numbers) or they may be functions (or "fields") which may be said to have an infinite number of degrees-of-freedom. If these variables are finite dimensional one talks of *discrete parameter systems* and typical examples of such systems are trusses, as shown in Figs. 1.2 and 1.4, where the state u is given by the collection of displacement vectors of nodes, and the design variable x is represented by a finite number of cross-sectional areas. On the other hand, if the design or state variable is a field, one talks of *distributed parameter systems* and such systems are, e.g., the shape optimization problem of Fig. 1.3 or the topology optimization problem in Fig. 1.5. Frequently in this text we use the term continuum problem for a distributed parameter system.

Now, distributed parameter systems are not suited for solution with a computer: computer implementations of mechanical problems are based on algebra, which is finite dimensional. This means that in the process of solving a distributed parameter system one performs a discretization, which produces a discrete parameter system. To distinguish between such derived discrete systems and systems like a truss structure we talk of *naturally* discrete parameter systems in the latter case. Ideally one would like to know that the discretized problem really is connected to the distributed one, i.e., one would like to prove that the solution of the discretized problem converges to the solution of the distributed one when the discretization is made finer and finer. However, such results are usually very mathematically demanding to obtain, and convergence results are not always available. The structural engineer then has to rely on intuition, that the discretized problem produces a result that is close to that of the distributed problem.

Chapter 2
Examples of Optimization of Discrete Parameter Systems

The following chapter gives some examples of the general optimization problem (SO) introduced in the previous chapter. They all concern the problem of finding the cross-sectional areas of bars or beams, i.e. they are sizing problems. The list of such examples is the following:

1. Minimization of the weight of a two-bar truss subject to stress constraints.
2. Minimization of the weight of a two-bar truss subject to stress and instability constraints.
3. Minimization of the weight of a two-bar truss subject to stress and displacement constraints.
4. Minimization of the weight of a two-beam cantilever subject to a displacement constraint.
5. Minimization of the weight of a three-bar truss subject to stress constraints.
6. Minimization of the weight of a three-bar truss subject to a stiffness constraint.

A simple example of combined shape and sizing optimization of a two-bar truss is given in Exercise 2.5. Despite their simplicity, it turns out that these problems display several general features of structural optimization problems.

The solution methods we will use in this chapter are of a very simple nature, and are applicable only when solving optimization problems with one or two design variables. Later, in Chaps. 3–5, we will study solution methods that are suitable for larger problems, and resolve some of the problems presented in this chapter.

2.1 Weight Minimization of a Two-Bar Truss Subject to Stress Constraints

Consider the two-bar truss shown in Fig. 2.1. The bars have the same length L and Young's modulus E. The force $F > 0$, and for the angle α we assume $0 \leq \alpha \leq 90°$. We are to minimize the weight under stress constraints. The design variables are the cross-sectional areas A_1 and A_2. The objective function, i.e., the total weight of the truss, becomes

$$f(A_1, A_2) = (A_1 + A_2)\rho L, \tag{2.1}$$

where ρ is the density of the material. It may be noted that this particular objective function does not depend on any state variables. As design constraints we prescribe that the cross-sectional areas must, for obvious physical reasons, be nonnegative, i.e.,

$$A_1 \geq 0, \qquad A_2 \geq 0. \tag{2.2}$$

P.W. Christensen, A. Klarbring, *An Introduction to Structural Optimization*,
© Springer Science + Business Media B.V. 2009

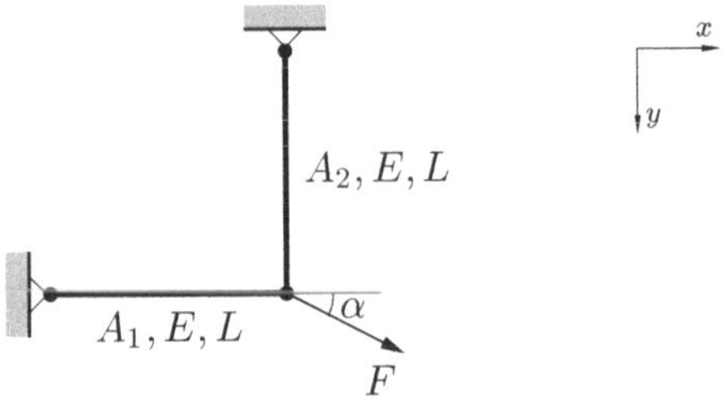

Fig. 2.1 Two-bar truss. Find the cross-sectional areas that minimize weight under stress constraints

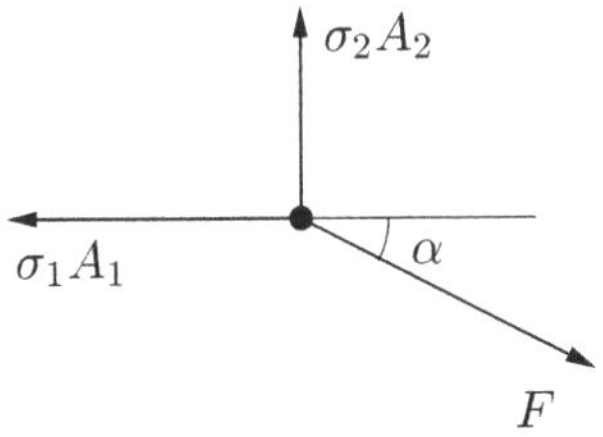

Fig. 2.2 Forces on the cut-out free node

In a truss problem of this type, the general approach would be to take the displacement vector $\boldsymbol{u}$ of the free node as state variable and then establish a state constraint of the form $\boldsymbol{K}(A_1, A_2)\boldsymbol{u} = \boldsymbol{F}$ by making use of all three basic conditions of small displacement elasticity theory, i.e. *equilibrium* in terms of forces and stresses, *geometric conditions* relating the bars' elongations to the displacement vector, and a linear *constitutive law*. However, in this particular problem the number of bars equals the number of degrees-of-freedom, which implies that the bar forces, or stresses, may be obtained directly from the equilibrium equations. We say that the truss is *statically determinate*. Furthermore, the displacement is not present in the constraints nor in the objective function. Therefore, we do not need to formulate any constitutive or geometric relations in order to write down the optimization problem. The equilibrium equations are found from the free-body diagram of the free node as shown in Fig. 2.2. Equilibrium in the x- and y-directions gives

$$F\cos\alpha - \sigma_1 A_1 = 0, \qquad F\sin\alpha - \sigma_2 A_2 = 0, \tag{2.3}$$

where we have opted to write the equations in terms of the bar stresses σ_1 and σ_2 directly, rather than first writing them in terms of the bar forces.

The state constraint involving stresses reads

$$|\sigma_i| \leq \sigma_0, \quad i = 1, 2, \tag{2.4}$$

where σ_0 is a maximum allowed stress level, the same in both tension and compression.

In summary, the particular version of the general ($\mathbb{SO}$) problem that is at hand here is to find A_1, A_2, σ_1 and σ_2 such that (2.1) is minimized under the constraints (2.2), (2.3) and (2.4). In a nested version of this problem we eliminate σ_1 and σ_2 by using (2.3) in (2.4) to find

$$-\sigma_0 A_1 \leq F\cos\alpha \leq \sigma_0 A_1,$$

$$-\sigma_0 A_2 \leq F \sin\alpha \leq \sigma_0 A_2.$$

Since $F, \cos\alpha, \sin\alpha, A_1, A_2 \geq 0$ it is clear that the left-hand inequalities in these expressions are always satisfied, i.e., they are redundant and can be left out of the problem. Furthermore, the right-hand inequalities are

$$A_1 \geq \frac{F\cos\alpha}{\sigma_0}, \qquad A_2 \geq \frac{F\sin\alpha}{\sigma_0},$$

which shows that the design constraints (2.2) are also redundant. We arrive at

$$(\mathbb{SO})^1_{\mathrm{nf}} \quad \begin{cases} \min\limits_{A_1, A_2} \; A_1 + A_2 \\ \text{s.t.} \; \begin{cases} A_1 \geq \dfrac{F\cos\alpha}{\sigma_0} \\ A_2 \geq \dfrac{F\sin\alpha}{\sigma_0}, \end{cases} \end{cases}$$

where the constant factor ρL has been left out of the objective function since it does not affect the optimum values of A_1 and A_2.

The problem $(\mathbb{SO})^1_{\mathrm{nf}}$ is a Linear Program (LP) in two variables and it is easily solved graphically as shown below. It should be noted that it is very unusual for a structural optimization problem to have a linear structure. In fact, it is even unusual for these problems to be convex. The fact that we find the LP structure in this case hinges on the simplicity of the constraints and objective function as well as on the statically determinate property.

In Fig. 2.3 a graphical solution of $(\mathbb{SO})^1_{\mathrm{nf}}$ is shown. In the A_1-A_2-plane we plot the lines defining the admissible domain. Next, we plot the line $A_1 + A_2 = \hat{f}(A_1, A_2) =$ constant, representing the objective function. The solution is found when $\hat{f}(A_1, A_2)$ is given the smallest possible value that maintains part of the line in the admissible region. It is given by

$$A_1^* = \frac{F\cos\alpha}{\sigma_0}, \qquad A_2^* = \frac{F\sin\alpha}{\sigma_0}.$$

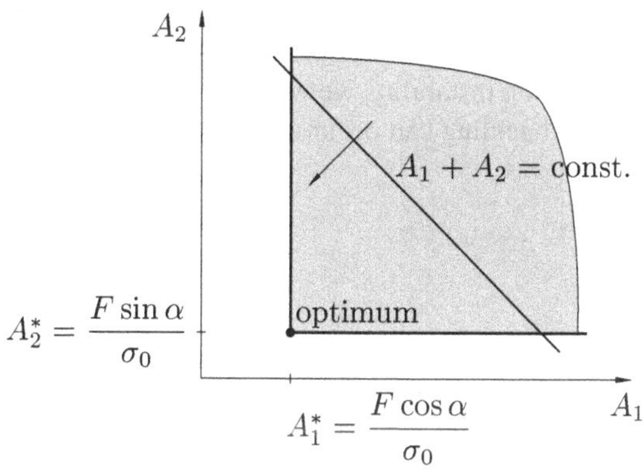

Fig. 2.3 Graphical solution of the problem

That is, both of the bars are fully used in tension: the stress is on the maximum level. It should be intuitively clear that this is a "good" structure from the point of view of using the least material.

Note that this problem, which is at the outset a sizing problem, is set so that topology may change: when $\alpha = 0$ or $90°$ one of the bars in the optimal solution "disappears."

2.2 Weight Minimization of a Two-Bar Truss Subject to Stress and Instability Constraints

Consider a two-bar truss consisting of bars of length L and Young's modulus E, placed at right angle according to Fig. 2.4. The force $F > 0$ is applied at an angle $\alpha = 45°$. The problem is to find the circular cross-sectional areas A_1 and A_2 such that the weight of the truss is minimized under constraints on stresses and Euler buckling. The weight of the truss is

$$f(A_1, A_2) = \rho L(A_1 + A_2),$$

where ρ is the density of the material. The stress constraints are as usual

$$|\sigma_i| \leq \sigma_0, \quad i = 1, 2, \tag{2.5}$$

where $\sigma_0 > 0$ is the stress bound. Equilibrium for the free node gives the stresses in the bars as

$$\sigma_1 = \frac{F}{\sqrt{2}A_1}, \qquad \sigma_2 = -\frac{F}{\sqrt{2}A_2},$$

so the stress constraints to be imposed in the optimization problem are

$$A_1 \geq \frac{F}{\sqrt{2}\sigma_0}, \qquad A_2 \geq \frac{F}{\sqrt{2}\sigma_0}. \tag{2.6}$$

Clearly, these constraints imply that cross-sectional areas will be nonnegative so we do not need to impose such restrictions explicitly.

Concerning instability, we want to obtain a safety factor of 4 against Euler buckling. Such buckling can occur only in the second bar, since there is tensile stress in

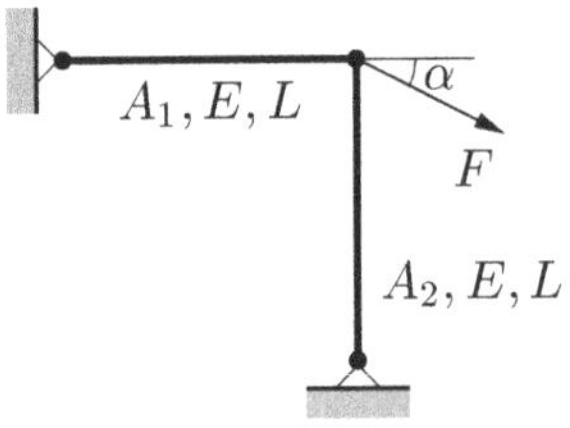

Fig. 2.4 A two-bar truss to be optimized under an instability constraint

the first bar. The buckling load for a hinged-hinged column is

$$P_c = \pi^2 \frac{EI}{L^2},$$

where for a circular cross section

$$I = \frac{A_2^2}{4\pi}.$$

Thus, the constraint

$$\frac{P_c}{4} \geq \sigma_2 A_2 = \frac{F}{\sqrt{2}}$$

becomes

$$A_2^2 \geq \frac{16FL^2}{\sqrt{2}\pi E}. \tag{2.7}$$

The optimization problem to be solved can, thus, be summarized as follows:

$$(\mathbb{SO})^2_{\text{nf}} \begin{cases} \min\limits_{A_1, A_2} \; A_1 + A_2 \\ \text{s.t.} \; \begin{cases} A_1 \geq \dfrac{F}{\sqrt{2}\sigma_0} \\ A_2 \geq \dfrac{F}{\sqrt{2}\sigma_0} \\ A_2^2 \geq \dfrac{16FL^2}{\sqrt{2}\pi E}. \end{cases} \end{cases}$$

Depending on the values of the coefficients, the second or the third constraint will be active. Consider, for instance, the special case

$$\sigma_0 = \frac{E}{100}, \qquad \sqrt{\frac{F}{\sigma_0}} = \frac{L}{4}.$$

Then, the constraints of $(\mathbb{SO})^2_{\text{nf}}$ become

$$A_1 \geq \frac{L^2}{16\sqrt{2}}, \qquad A_2 \geq \frac{L^2}{16\sqrt{2}}, \qquad A_2 \geq \frac{L^2}{10\sqrt{\sqrt{2}\pi}}$$

and since

$$1.6\sqrt{\sqrt{2}} > \sqrt{\pi} \quad \Leftrightarrow \quad \frac{L^2}{10\sqrt{\sqrt{2}\pi}} > \frac{L^2}{16\sqrt{2}},$$

it can be concluded that the optimum occurs when both the first and the third constraints are active, i.e., when

$$A_1^* = \frac{L^2}{16\sqrt{2}} \approx 0.044L^2, \qquad A_2^* = \frac{L^2}{10\sqrt{\sqrt{2}\pi}} \approx 0.047L^2.$$

2.3 Weight Minimization of a Two-Bar Truss Subject to Stress and Displacement Constraints

Consider the truss in Fig. 2.5. The bars have lengths according to the figure, and consist of a material with Young's modulus E and density ρ. The force $F > 0$ and the angle $\alpha = 30°$. We want to find the cross-sectional areas A_1 and A_2 such that the weight is minimized subject to stress constraints and a constraint on the tip displacement δ. The weight can be written

$$f(A_1, A_2) = \rho L \left(\frac{2}{\sqrt{3}} A_1 + A_2 \right). \tag{2.8}$$

The stress constraints are

$$|\sigma_i| \leq \sigma_0, \quad i = 1, 2, 3, \tag{2.9}$$

for a given stress bound $\sigma_0 > 0$. The displacement constraint is

$$\delta \leq \delta_0, \tag{2.10}$$

where

$$\delta_0 = \frac{\sigma_0 L}{E},$$

is a given bound on the tip displacement. The design constraints are

$$A_1 \geq 0, \qquad A_2 \geq 0. \tag{2.11}$$

We are aiming at a nested formulation, and need to rewrite (2.9) and (2.10) in terms of cross-sectional areas. Equilibrium equations are obtained from Fig. 2.6.

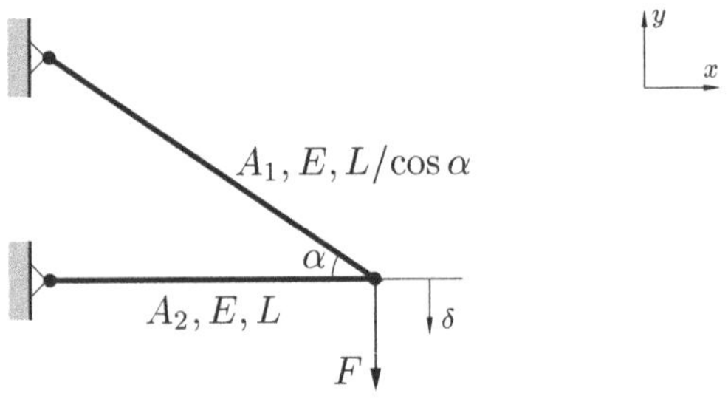

Fig. 2.5 Two bar truss subject to stress and displacement constraints

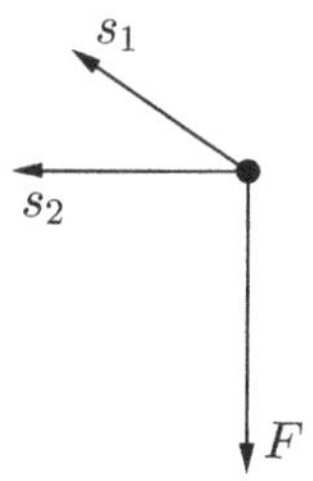

Fig. 2.6 Forces on the cut-out free node

The equations for the x- and y-directions become

$$-s_1 \cos\alpha - s_2 + F_x = 0, \qquad s_1 \sin\alpha + F_y = 0,$$

where s_1 and s_2 are the bar forces, $F_x = 0$ and $F_y = -F$. These equations may be written in matrix form as

$$\begin{bmatrix} F_x \\ F_y \end{bmatrix} = \begin{bmatrix} \frac{\sqrt{3}}{2} & 1 \\ -\frac{1}{2} & 0 \end{bmatrix} \begin{bmatrix} s_1 \\ s_2 \end{bmatrix}. \tag{2.12}$$

In symbolic matrix form this is written $\boldsymbol{F} = \boldsymbol{B}^T \boldsymbol{s}$. Here, superscript T denotes the transpose of a matrix; it will soon become apparent why we write (2.12) symbolically by use of the transpose of a matrix.

Since the number of bars equals the number of degrees-of-freedom, the truss is statically determinate, and we may obtain the bar forces $\boldsymbol{s}$ by simply solving the equilibrium equations. From (2.12) we get

$$\boldsymbol{s} = \begin{bmatrix} s_1 \\ s_2 \end{bmatrix} = \boldsymbol{B}^{-T} \boldsymbol{F} = \begin{bmatrix} 2F \\ -\sqrt{3}F \end{bmatrix}. \tag{2.13}$$

In order to rewrite the displacement constraint (2.10) in terms of cross-sectional areas, we need to include geometric and constitutive conditions. In a small displacement theory, the elongations of the bars, δ_1 and δ_2, are obtained by projecting the displacement vector $\boldsymbol{u} = [(u_x \ u_y)]^T$ of the free node on the unit vectors directed along the bars and pointing towards the free node:

$$\boldsymbol{e}_1 = \begin{bmatrix} \frac{\sqrt{3}}{2} \\ -\frac{1}{2} \end{bmatrix}, \qquad \boldsymbol{e}_2 = \begin{bmatrix} 1 \\ 0 \end{bmatrix}.$$

The elongations thus become

$$\delta_1 = \boldsymbol{e}_1^T \boldsymbol{u} = \frac{\sqrt{3}}{2} u_x - \frac{1}{2} u_y, \qquad \delta_2 = \boldsymbol{e}_2^T \boldsymbol{u} = u_x.$$

In matrix form this reads

$$\begin{bmatrix} \delta_1 \\ \delta_2 \end{bmatrix} = \begin{bmatrix} \frac{\sqrt{3}}{2} & -\frac{1}{2} \\ 1 & 0 \end{bmatrix} \begin{bmatrix} u_x \\ u_y \end{bmatrix}. \tag{2.14}$$

The—perhaps surprising—fact that occurs here is that the matrix of this equation is $\boldsymbol{B}$, i.e., the transpose of the matrix occurring in (2.12), so (2.14) can in symbolic matrix form be written as $\boldsymbol{\delta} = \boldsymbol{B}\boldsymbol{u}$. That $\boldsymbol{B}^T$ and $\boldsymbol{B}$ appear in this way in the equilibrium and geometric equations is not a coincidence: the same property holds in any truss problem and, in fact, given the right interpretation, in any small displacement structural problem. It is related to the validity of the work equation $\boldsymbol{\delta}^T \boldsymbol{s} = \boldsymbol{u}^T \boldsymbol{F}$, and it is a very economical fact since, given equilibrium, we can directly write down the geometric equations and vice versa.

Next, we need the constitutive equations. Hooke's law reads $\sigma_i = E\varepsilon_i$, where

$$\sigma_i = s_i/A_i, \qquad \varepsilon_i = \delta_i/l_i,$$

are the stress and strain in bar i of length l_i. Combining these equations gives us the elongations in terms of the bar forces as

$$\delta_i = \frac{l_i s_i}{A_i E}. \tag{2.15}$$

From (2.13), and since $l_1 = 2L/\sqrt{3}$ and $l_2 = L$, we get

$$\boldsymbol{\delta} = \begin{bmatrix} \delta_1 \\ \delta_2 \end{bmatrix} = \begin{bmatrix} \dfrac{4FL}{\sqrt{3}A_1 E} \\ -\dfrac{\sqrt{3}FL}{A_2 E} \end{bmatrix}.$$

The displacements of the free node are thus given by

$$\boldsymbol{u} = \boldsymbol{B}^{-1}\boldsymbol{\delta} = \frac{FL}{E}\begin{bmatrix} -\dfrac{\sqrt{3}}{A_2} \\ -\dfrac{8}{\sqrt{3}A_1} - \dfrac{3}{A_2} \end{bmatrix}.$$

The tip displacement may now be written in terms of the cross-sectional areas as

$$\delta = -\boldsymbol{e}_y^T \boldsymbol{u} = \frac{FL}{E}\left(\frac{8}{\sqrt{3}A_1} + \frac{3}{A_2}\right),$$

where $\boldsymbol{e}_y$ is the unit vector in the y-direction, so that (2.10) can be written

$$\frac{8}{\sqrt{3}A_1} + \frac{3}{A_2} \leq \frac{E\delta_0}{FL} = \frac{\sigma_0}{F}. \tag{2.16}$$

Regarding the stress constraints, we note from (2.13) and $F > 0$ that bar 1 is in tension and bar 2 in compression, so we need to consider only the stress constraints $s_1/A_1 \leq \sigma_0$ and $-s_2/A_2 \leq \sigma_0$, which with (2.13) lead to

$$A_1 \geq \frac{2F}{\sigma_0}, \qquad A_2 \geq \frac{\sqrt{3}F}{\sigma_0}. \tag{2.17}$$

Since $F > 0$ and $\sigma_0 > 0$, we conclude that (2.11) are redundant: the optimal cross-sectional areas are strictly positive.

In summary, our problem is to minimize $f(A_1, A_2)$, according to (2.8), under constraints given by (2.16) and (2.17). Now, we will not treat this problem directly, but instead rewrite the problem by means of a change of variables. We do this to demonstrate the use of such ideas, since they will play an essential role in upcoming sections, and one may consider that the problem is also easier to solve in the new variables. These new, dimensionless variables are

$$x_1 = \frac{2F}{\sigma_0 A_1} > 0, \qquad x_2 = \frac{\sqrt{3}F}{\sigma_0 A_2} > 0,$$

and the essential thing with these new variables is that they make the constraint (2.16) linear. Moreover, the new variables are scaled such that (2.17) becomes

$$1 \geq x_1, \qquad 1 \geq x_2. \tag{2.18}$$

The displacement constraint (2.16) now becomes

$$\frac{4}{\sqrt{3}}x_1 + \sqrt{3}x_2 \leq 1, \tag{2.19}$$

and the objective function (2.8) is written as

$$f(A_1(x_1), A_2(x_2)) = \frac{\sqrt{3}\rho L F}{\sigma_0}\left(\frac{4}{3x_1} + \frac{1}{x_2}\right). \tag{2.20}$$

The constraint (2.19) gives the following estimates

$$1 \geq \frac{4}{\sqrt{3}}x_1 + \sqrt{3}x_2 \geq \sqrt{3}x_2, \qquad 1 \geq \frac{4}{\sqrt{3}}x_1 + \sqrt{3}x_2 \geq \frac{4}{\sqrt{3}}x_1,$$

from which it is clear that (2.18) is redundant.

We have arrived at the following optimization problem

$$(\mathbb{SO})^3_{\text{nf}} \quad \begin{cases} \min\limits_{x_1, x_2} \ \hat{f}(x_1, x_2) = \dfrac{4}{3x_1} + \dfrac{1}{x_2} \\ \text{s.t.} \quad \begin{cases} \dfrac{4}{\sqrt{3}}x_1 + \sqrt{3}x_2 \leq 1 \\ x_1 > 0, \qquad x_2 > 0. \end{cases} \end{cases}$$

This problem is illustrated in Fig. 2.7, from which we conclude that constraint (2.19) is active. We write (2.19) as an equality and solve for x_2 to obtain

$$x_2 = \frac{1}{\sqrt{3}} - \frac{4}{3}x_1.$$

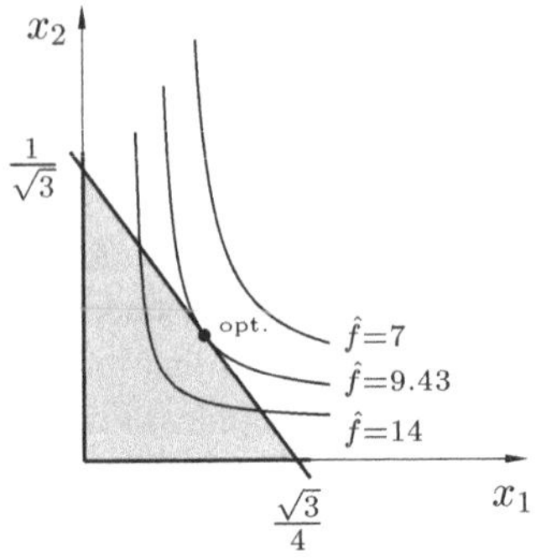

Fig. 2.7 Geometric illustration of $(\mathbb{SO})^3_{\text{nf}}$

This is then substituted into $\hat{f}(x_1, x_2)$, which becomes a function of x_1 for which we seek a stationary value. Such a stationary value is realized whenever

$$x_1 = \pm\left(\frac{1}{\sqrt{3}} - \frac{4}{3}x_1\right).$$

The minus sign gives $x_1 = \sqrt{3}$ which is greater than 1 and, thus, not in the admissible domain. The plus sign gives the solution

$$x_1^* = \frac{\sqrt{3}}{7}, \qquad x_2^* = \frac{\sqrt{3}}{7},$$

which, going back to the original variables, gives

$$A_1^* = \frac{14F}{\sqrt{3}\sigma_0} \approx \frac{8.1F}{\sigma_0}, \qquad A_2^* = \frac{7F}{\sigma_0}.$$

2.4 Weight Minimization of a Two-Beam Cantilever Subject to a Displacement Constraint

Consider a cantilever beam, fixed at the left end and subject to a vertical force $F > 0$ at the right end. The beam consists of N segments, each of length L, so the total length of the cantilever is NL. Segment number N is to the left, at the built-in end, and segment 1 is at the free end. Each segment cross section has a hollow square form, see Fig. 2.8. The thickness of the material is t for all segments, and the length of the side of the square is x_A for segment $A = 1, \ldots, N$. The bending moment of inertia, I_A can be calculated from classical formulas. If it is assumed that $t \ll x_A$, for all A, one finds:

$$I_A = \frac{x_A^4}{12} - \frac{(x_A - 2t)^4}{12} = \frac{2tx_A^3}{3}.$$

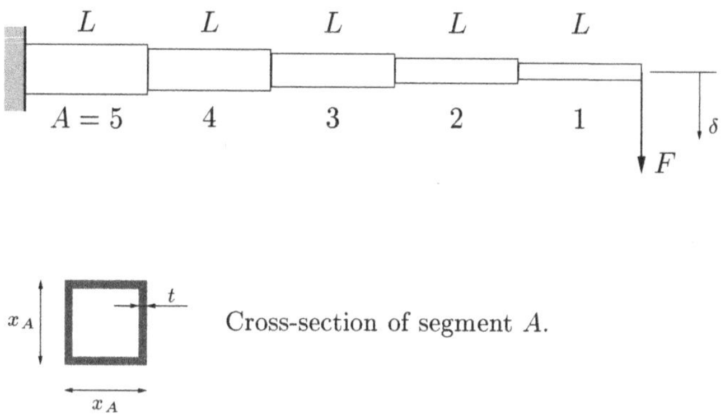

Fig. 2.8 The cantilever for 5 segments and the hollow square cross section

We want to minimize the weight of the beam under the constraint that the displacement at the tip, δ, is less than some prescribed value δ_0. The design variables are the cross-sectional sizes x_A, $A = 1, \ldots, N$. The weight when $t \ll x_A$ becomes

$$f(x_1, \ldots, x_N) = L\rho \sum_{A=1}^{N} \left(x_A^2 - (x_A - 2t)^2 \right) = 4L\rho t \sum_{A=1}^{N} x_A,$$

where ρ is the density. The displacement at the tip of the beam can be seen as the sum of contributions from each segment, when other segments are considered as rigid, i.e.,

$$\delta = \sum_{A=1}^{N} \delta^{(A)}, \tag{2.21}$$

where $\delta^{(A)}$ is the displacement at the tip of the cantilever for a system where only segment A is elastic. Next, one concludes by simple geometry for small displacements, such that $\sin\theta_A \approx \theta_A$, that

$$\delta^{(A)} = \delta_A + (A-1)L\theta_A, \tag{2.22}$$

where δ_A and θ_A are the displacement and the rotation at the right-hand side of segment A when only this segment is elastic, see Fig. 2.9. One calculates δ_A and θ_A by means of elementary beam theory as follows:

$$\delta_A = \frac{M_A L^2}{2EI_A} + \frac{F_A L^3}{3EI_A}, \tag{2.23}$$

$$\theta_A = \frac{M_A L}{EI_A} + \frac{F_A L^2}{2EI_A}, \tag{2.24}$$

where E is Young's modulus, and

$$M_A = (A-1)LF, \qquad F_A = F,$$

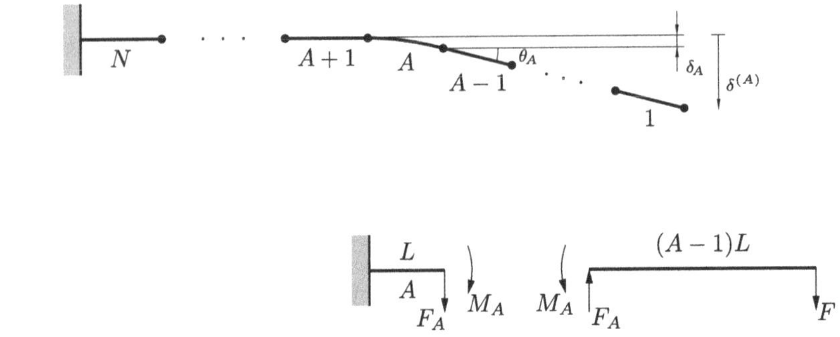

Fig. 2.9 The cantilever when only segment A is elastic

are the bending moment and the shear force at the right end of segment A. Putting (2.23) and (2.24) into (2.22), the result into (2.21) and using the above expression for I_A gives

$$\delta = \frac{3FL^3}{2Et} \sum_{A=1}^{N} \left(A^2 - A + \frac{1}{3} \right) \frac{1}{x_A^3}. \tag{2.25}$$

The present cantilever problem was originally formulated and solved analytically as well as numerically in Svanberg [34] for the case $N = 5$. Here we will be content with $N = 2$, which is easily solved analytically. For this case we have the following optimization problem:

$$(\mathbb{SO})_{\text{nf}}^4 \quad \begin{cases} \min\limits_{x_1, x_2} & f(x_1, x_2) = C_1(x_1 + x_2) \\ \text{s.t.} & \begin{cases} \dfrac{1}{x_1^3} + \dfrac{7}{x_2^3} \leq C_2 \\ x_1 > 0, \qquad x_2 > 0, \end{cases} \end{cases}$$

where

$$C_1 = 4\rho L t, \qquad C_2 = \frac{2\delta_0 E t}{FL^3}.$$

Assuming equality in the nonstrict inequality constraint we solve this for x_2. The result is put into $f(x_1, x_2)$, which becomes a function of x_1 only. Seeking a stationary value of this function gives the solution

$$x_1^* = \left(\frac{1 + 7^{1/4}}{C_2} \right)^{1/3}, \qquad x_2^* = 7^{1/4} \left(\frac{1 + 7^{1/4}}{C_2} \right)^{1/3}.$$

Now, one may ask the question, what happens if we reverse the order of the structural measures in a problem of this kind, i.e., what if we minimize the tip displacement under a constraint on the weight? We then have the following problem:

$$\begin{cases} \min\limits_{x_1,x_2} \dfrac{1}{x_1^3} + \dfrac{7}{x_2^3} \\ \text{s.t.} \quad \begin{cases} C_1(x_1+x_2) \le W \\ x_1 > 0, \ x_2 > 0, \end{cases} \end{cases}$$

where W is some given allowable weight. This problem can be solved in the same way as $(\mathbb{SO})^4_{\mathrm{nf}}$. One finds the solution

$$x_1^{**} = \frac{W}{C_1}\left(\frac{1}{1+7^{1/4}}\right), \qquad x_2^{**} = \frac{W}{C_1}\left(\frac{7^{1/4}}{1+7^{1/4}}\right),$$

and it can be concluded that the reversed problem gives a solution different from $(\mathbb{SO})^4_{\mathrm{nf}}$. However, it holds that

$$\frac{x_2^*}{x_1^*} = \frac{x_2^{**}}{x_1^{**}} = 7^{1/4} \approx 1.63.$$

Thus, the solution of $(\mathbb{SO})^4_{\mathrm{nf}}$ can be obtained by a scaling of the solution of the reversed problem and vice versa. This is a general property which will be discussed more thoroughly in Sect. 5.2.3.

2.5 Weight Minimization of a Three-Bar Truss Subject to Stress Constraints

Consider the three-bar truss shown in Fig. 2.10. The bars have Young's modulus E and the lengths are $l_1 = L$, $l_2 = L$, $l_3 = L/\beta$, where $\beta > 0$. In this example $\beta = 1$, but in the next section the same truss will be studied with $\beta = 1/10$. We will therefore perform all derivations for a general $\beta > 0$. The force $F > 0$. As for the two-bar truss in Sect. 2.1 we are to minimize the weight under stress constraints. The design variables are the cross-sectional areas A_1, A_2 and A_3, but for simplicity

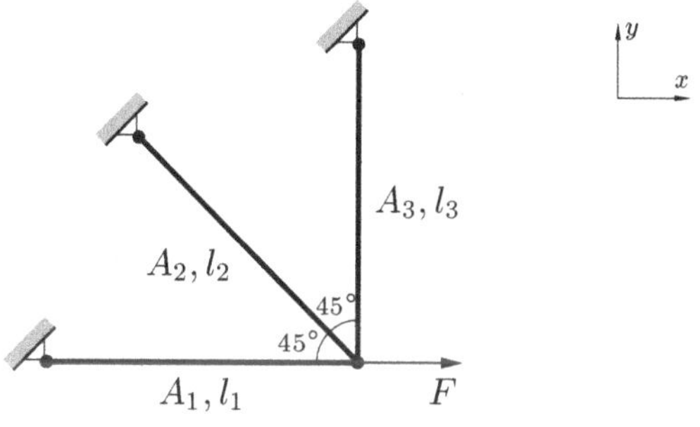

Fig. 2.10 Three-bar truss. Find the cross-sectional areas that minimize weight under stress constraints

we assume that

$$A_1 = A_3.$$

The objective function, which is the total weight of the truss, becomes

$$f(A_1, A_2) = \rho_1 L A_1 + \rho_2 L A_2 + \rho_3 \frac{L}{\beta} A_3 = L\left(\rho_1 + \frac{\rho_3}{\beta}\right) A_1 + L\rho_2 A_2, \quad (2.26)$$

where ρ_1, ρ_2 and ρ_3 are the densities of the bars. The design constraints are

$$A_1 \geq 0, \qquad A_2 \geq 0. \quad (2.27)$$

Concerning designs with A_1 or A_2 equal to zero, it is clearly impossible to have $A_1 = A_3 = 0$ since then there is no equilibrium possible as it would imply collapse of the structure under the given external load. On the other hand, $A_2 = 0$ is a valid design.

The state constraints are that the maximum absolute value of the stress in bar i must not exceed the values $\sigma_i^{\max}$, i.e.

$$|\sigma_i| \leq \sigma_i^{\max}, \quad i = 1, 2, 3. \quad (2.28)$$

The equilibrium equation is found by cutting out the free node as shown in Fig. 2.11. The equilibrium equations in the x- and y-directions become

$$-s_1 - \frac{s_2}{\sqrt{2}} + F = 0, \qquad s_3 + \frac{s_2}{\sqrt{2}} = 0.$$

In matrix form these equations read

$$\begin{bmatrix} F \\ 0 \end{bmatrix} = \begin{bmatrix} 1 & \frac{1}{\sqrt{2}} & 0 \\ 0 & -\frac{1}{\sqrt{2}} & -1 \end{bmatrix} \begin{bmatrix} s_1 \\ s_2 \\ s_3 \end{bmatrix} \quad \Longleftrightarrow \quad \boldsymbol{F} = \boldsymbol{B}^T \boldsymbol{s}. \quad (2.29)$$

Note that in contrast to the two-bar truss in Sect. 2.3, we cannot obtain the bar forces from the equilibrium equations alone since the number of bars exceeds the number of degrees-of-freedom. We say that the truss is *statically indeterminate*. In order to find the bar forces, or, rather, the stresses, that appear in the constraints, we need to make use of Hooke's law and the geometry conditions.

From (2.15) we have

$$s_i = \frac{E A_i \delta_i}{l_i}.$$

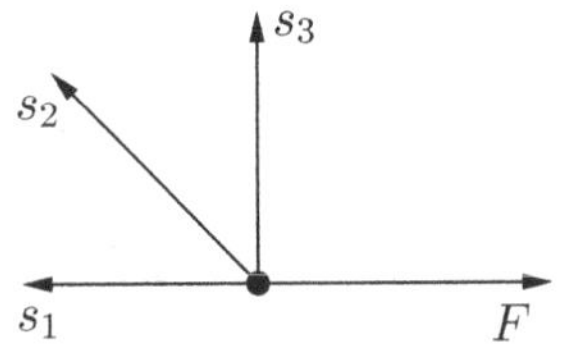

Fig. 2.11 Forces on the cut-out free node

We write these equations for all three bars in matrix form as

$$s = D\delta,$$

where

$$D = \frac{E}{l}\begin{bmatrix} A_1 & 0 & 0 \\ 0 & A_2 & 0 \\ 0 & 0 & \beta A_1 \end{bmatrix}.$$

Since $\delta = Bu$, cf. the discussion following (2.14), the bar forces are obtained as

$$s = DBu. \tag{2.30}$$

The equilibrium equations (2.29) thus become

$$F = B^T s = B^T DBu = Ku, \tag{2.31}$$

where $K = B^T DB$ is the global stiffness matrix of the truss, which is easily calculated as

$$K = \frac{E}{l}\begin{bmatrix} A_1 + \dfrac{A_2}{2} & -\dfrac{A_2}{2} \\ -\dfrac{A_2}{2} & \dfrac{A_2}{2} + \beta A_1 \end{bmatrix}.$$

From (2.31) we obtain the displacements of the free node as $u = K^{-1}F$:

$$u_x = \frac{FL}{EA_1}\left(\frac{2\beta A_1 + A_2}{2\beta A_1 + (1+\beta)A_2}\right), \tag{2.32}$$

$$u_y = \frac{FL}{EA_1}\left(\frac{A_2}{2\beta A_1 + (1+\beta)A_2}\right). \tag{2.33}$$

Using (2.30), the stresses may be written as

$$\sigma = As = ADBu,$$

where

$$A = \begin{bmatrix} \dfrac{1}{A_1} & 0 & 0 \\ 0 & \dfrac{1}{A_2} & 0 \\ 0 & 0 & \dfrac{1}{A_1} \end{bmatrix}.$$

Some straightforward calculations give us the bar stresses as

$$\sigma_1 = \frac{F}{2\beta A_1 + (1+\beta)A_2}\left(2\beta + \frac{A_2}{A_1}\right), \tag{2.34}$$

$$\sigma_2 = \frac{\sqrt{2}F\beta}{2\beta A_1 + (1+\beta)A_2}, \tag{2.35}$$

$$\sigma_3 = -\frac{F\beta \dfrac{A_2}{A_1}}{2\beta A_1 + (1+\beta)A_2}. \tag{2.36}$$

Since $F, A_1, A_2 > 0$, we conclude that bars 1 and 2 are in tension and bar 3 is in compression, so only the stress constraints $\sigma_1 \leq \sigma_1^{\max}$, $\sigma_2 \leq \sigma_2^{\max}$ and $-\sigma_3^{\max} \leq \sigma_3$ need to be considered. In what follows we will put $\beta = 1$, i.e. $l_3 = L$. The stress constraint $\sigma_1 \leq \sigma_1^{\max}$ then takes the form

$$\frac{F(2A_1 + A_2)}{2A_1(A_1 + A_2)} \leq \sigma_1^{\max}. \tag{2.37}$$

The constraint $\sigma_2 \leq \sigma_2^{\max}$ reads

$$\frac{F}{\sqrt{2}(A_1 + A_2)} \leq \sigma_2^{\max}. \tag{2.38}$$

Naturally, this constraint should only be included if bar 2 is present, i.e. if $A_2 > 0$. Finally, the stress constraint $-\sigma_3^{\max} \leq \sigma_3$ is written as

$$\frac{FA_2}{2A_1(A_1 + A_2)} \leq \sigma_3^{\max}. \tag{2.39}$$

We have arrived at the following optimization problem

$$(\mathbb{SO})_{\mathrm{nf}}^{5} \quad \begin{cases} \min\limits_{A_1, A_2} \ (\rho_1 A_1 + \rho_2 A_2 + \rho_3 A_1)\, L \\ \text{s.t.} \ \begin{cases} \dfrac{F(2A_1 + A_2)}{2A_1(A_1 + A_2)} - \sigma_1^{\max} \leq 0 \\ \dfrac{F}{\sqrt{2}(A_1 + A_2)} - \sigma_2^{\max} \leq 0 \quad \text{if } A_2 > 0 \\ \dfrac{FA_2}{2A_1(A_1 + A_2)} - \sigma_3^{\max} \leq 0 \\ A_1 > 0, \qquad A_2 \geq 0. \end{cases} \end{cases}$$

In order to illustrate that all bars may not be present in the optimal truss, and that structural optimization problems may have more than one, and even an infinite number of solutions, we will solve this problem for five different cases by altering the density and the yield stress of the bars.

CASE A)
$\rho_1 = 2\rho_0, \ \rho_2 = \rho_3 = \rho_0, \ \sigma_1^{\max} = \sigma_2^{\max} = \sigma_3^{\max} = \sigma_0.$
By introducing the new dimensionless variables x_1 and x_2 as

$$x_1 = \frac{A_1\sigma_0}{F}, \qquad x_2 = \frac{A_2\sigma_0}{F}, \tag{2.40}$$

we may write the optimization problem as

$$
(\mathbb{SO})_{\text{nf}}^{5a} \quad \begin{cases} \min\limits_{x_1,x_2} \; 3x_1 + x_2 \\ \text{s.t.} \quad \begin{cases} \dfrac{2x_1 + x_2}{2x_1(x_1 + x_2)} - 1 \leq 0 & & (\sigma_1) \\ \dfrac{1}{\sqrt{2}(x_1 + x_2)} - 1 \leq 0 & \text{if } x_2 > 0 & (\sigma_2) \\ \dfrac{x_2}{2x_1(x_1 + x_2)} - 1 \leq 0 & & (\sigma_3) \\ x_1 > 0, \qquad x_2 \geq 0, \end{cases} \end{cases}
$$

where, for simplicity, the objective function is the weight divided by the positive scalar $FL\rho_0/\sigma_0$. The problem is illustrated in Fig. 2.12. Note that the σ_2-constraint is linear. It is clear that the σ_1-constraint (2.37) is active at the solution, and that all other constraints are inactive. Thus

$$2x_1 + x_2 - 2x_1(x_1 + x_2) = 0,$$

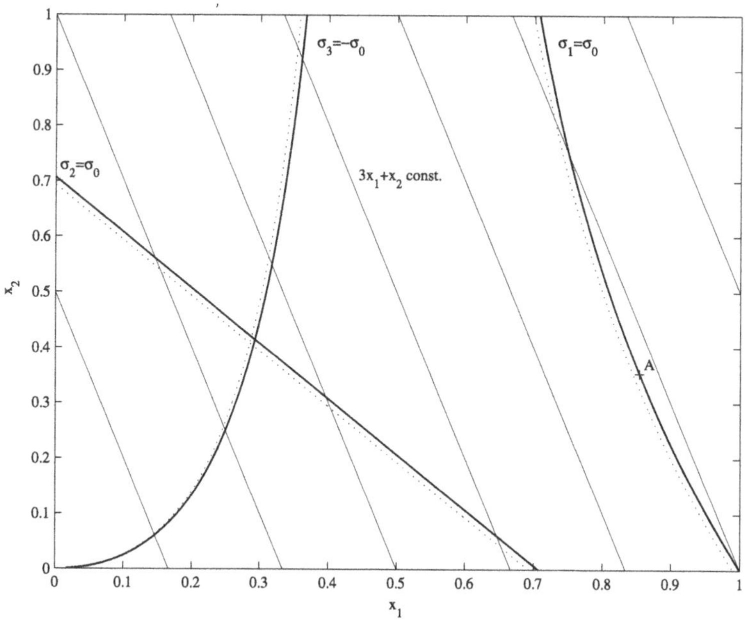

Fig. 2.12 Case a). A *solid thick line* with a *dotted line* alongside indicates a constraint; the region on the same side of the *thick line* as the corresponding *dotted line* is not part of the design space. The *thin solid lines* are iso-merit lines, i.e. all points on a *thin line* yield the same value of the objective function. Point A is the solution

which gives

$$x_2 = \frac{2x_1(x_1-1)}{1-2x_1}. \tag{2.41}$$

Substituting this into the objective function we find that the problem is reduced to minimizing

$$3x_1 + \frac{2x_1(x_1-1)}{1-2x_1},$$

for $x_1 > 0$. We find that this function has a stationary value for x_1 satisfying the second order equation

$$8x_1^2 - 8x_1 + 1 = 0.$$

The solution of this equation is

$$x_1^* = \frac{1}{2} \pm \frac{\sqrt{2}}{4},$$

where the minus sign is not valid since upon substitution into (2.41) it gives a negative x_2^*. Using the plus sign instead, gives

$$x_2^* = \frac{\sqrt{2}}{4}.$$

Reverting to the original area variables A_1 and A_2, cf. (2.40), the optimal solution is

$$A_1^* = \frac{F}{2\sigma_0}\left(1 + \frac{1}{\sqrt{2}}\right), \qquad A_2^* = \frac{F}{2\sqrt{2}\sigma_0},$$

and the corresponding optimum weight is

$$(3A_1^* + A_2^*)\rho_0 L = \frac{FL\rho_0}{\sigma_0}\left(\frac{3}{2} + \sqrt{2}\right).$$

CASE B)
$\rho_1 = \rho_2 = \rho_3 = \rho_0,\ \sigma_1^{\max} = \sigma_3^{\max} = 2\sigma_0,\ \sigma_2^{\max} = \sigma_0.$
Using the same dimensionless variables as for the previous case, we may write the problem as

$$(\mathbb{SO})_{\text{nf}}^{5b} \quad \begin{cases} \min\limits_{x_1,x_2} \ 2x_1 + x_2 \\ \text{s.t.} \ \begin{cases} \dfrac{2x_1+x_2}{4x_1(x_1+x_2)} - 1 \le 0 & & (\sigma_1) \\ \dfrac{1}{\sqrt{2}(x_1+x_2)} - 1 \le 0 & \text{if } x_2 > 0 & (\sigma_2) \\ \dfrac{x_2}{4x_1(x_1+x_2)} - 1 \le 0 & & (\sigma_3) \\ x_1 > 0, \quad x_2 \ge 0, \end{cases} \end{cases}$$

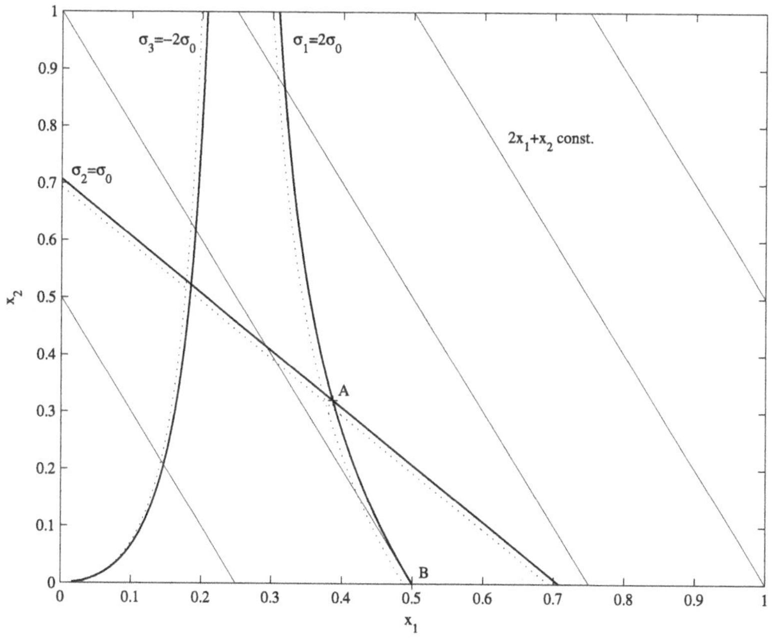

Fig. 2.13 Case b). Point B is the solution

see Fig. 2.13. It would appear that the solution is at the intersection A of the σ_1- and σ_2-constraints. However, we must keep in mind that the σ_2-constraint is valid only for $x_2 > 0$. By deleting this constraint, it is evident from the figure, that the point B on the σ_1-constraint curve, where x_2 is zero, gives the lowest weight that can be attained. This point is obtained by letting $x_2 = 0$ in the active σ_1-constraint:

$$2x_1 - 4x_1^2 = 0,$$

which gives $x_1^* = 1/2$ as $x_1^* = 0$ is not a valid design. In the original variables, the optimum solution becomes

$$A_1^* = \frac{F}{2\sigma_0}, \qquad A_2^* = 0,$$

with the optimal weight

$$\frac{FL\rho_0}{\sigma_0}.$$

CASE C)
$\rho_1 = (2\sqrt{2}-1)\rho_0,\ \rho_2 = \rho_3 = \rho_0,\ \sigma_1^{\max} = \sigma_3^{\max} = 2\sigma_0,\ \sigma_2^{\max} = \sigma_0.$

The density of bar 1 is now increased somewhat as compared to case b). This will alter the objective function but not the constraints:

$$(\mathbb{SO})_{\text{nf}}^{5c} \quad \begin{cases} \min\limits_{x_1,x_2} \; 2\sqrt{2}x_1 + x_2 \\ \text{s.t. the constraints in } (\mathbb{SO})_{\text{nf}}^{5b}, \end{cases}$$

which is illustrated in Fig. 2.14. It is not evident from the figure whether the solution is at the intersection A between the σ_1- and σ_2-constraints, or the point B corresponding to a design without bar 2. Point A may be calculated by solving the two equations obtained when equality is satisfied in the σ_1- and σ_2-constraints, which leads to

$$x_1^* = \frac{4+\sqrt{2}}{14}, \qquad x_2^* = \frac{6\sqrt{2}-4}{14}.$$

Point B is $x_1^{**} = 1/2$, $x_2^{**} = 0$. It turns out that these two points yield the same value of the objective function, and thus, there are two solutions to this problem! In the original variables, the solutions are written

$$A_1^* = \frac{F}{\sigma_0}\left(\frac{4+\sqrt{2}}{14}\right), \qquad A_2^* = \frac{F}{\sigma_0}\left(\frac{6\sqrt{2}-4}{14}\right),$$

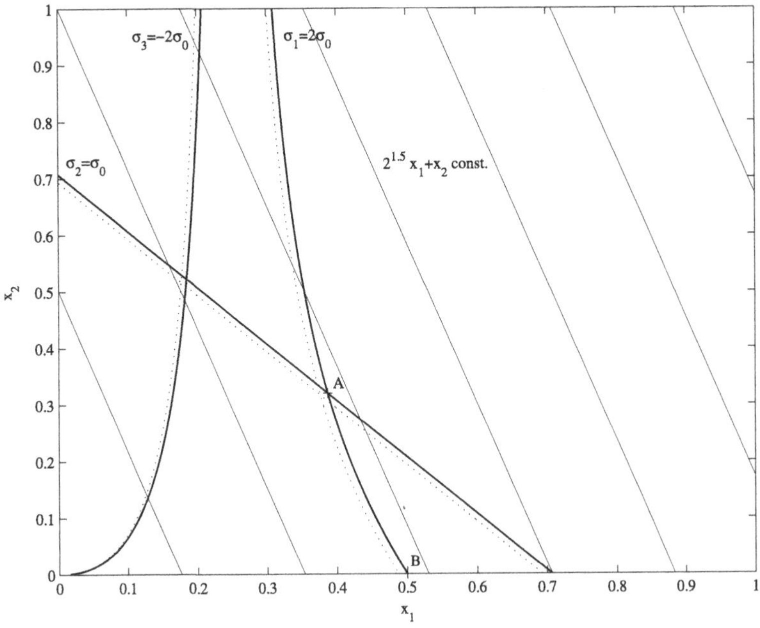

Fig. 2.14 Case c). Points A and B are the solutions

$$A_1^{**} = \frac{F}{2\sigma_0}, \qquad A_2^{**} = 0,$$

with the optimum weight

$$\frac{\sqrt{2}FL\rho_0}{\sigma_0}.$$

CASE D)
$\rho_1 = 3\rho_0,\ \rho_2 = \rho_3 = \rho_0,\ \sigma_1^{\max} = \sigma_3^{\max} = 2\sigma_0,\ \sigma_2^{\max} = \sigma_0.$
Again, the density of bar 1 is increased. The optimization problem becomes

$$(\mathbb{SO})_{\text{nf}}^{5d} \quad \begin{cases} \min\limits_{x_1,x_2} \ 4x_1 + x_2 \\ \text{s.t. the constraints in } (\mathbb{SO})_{\text{nf}}^{5b}. \end{cases}$$

In Fig. 2.15, we see that the σ_1- and σ_2-constraints are active at the solution. This point has already been calculated for case c) as

$$A_1^* = \frac{F}{\sigma_0}\left(\frac{4+\sqrt{2}}{14}\right), \qquad A_2^* = \frac{F}{\sigma_0}\left(\frac{6\sqrt{2}-4}{14}\right),$$

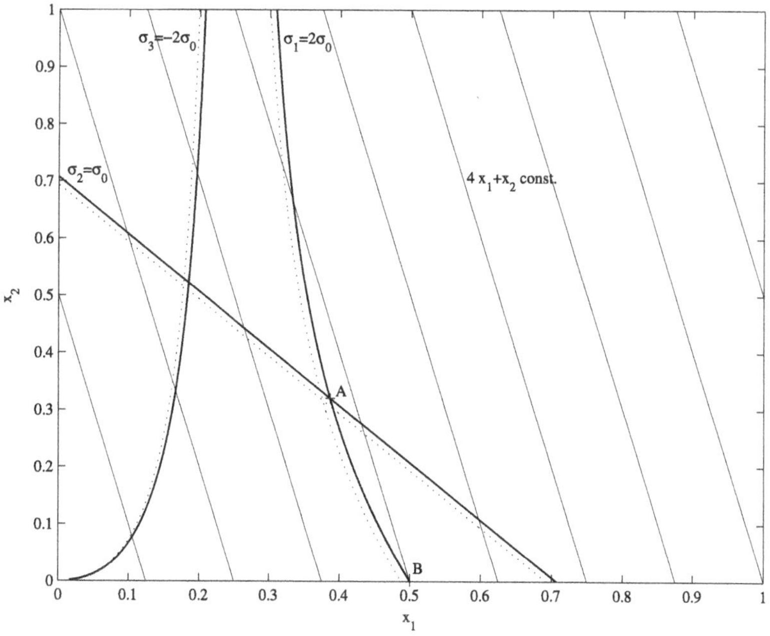

Fig. 2.15 Case d). Point A is the solution

which gives the optimal weight

$$\frac{FL\rho_0}{\sigma_0}\left(\frac{6+5\sqrt{2}}{7}\right).$$

CASE E)

$\rho_1 = \rho_3 = \rho_0,\ \rho_2 = 2\rho_0,\ \sigma_1^{\max} = \sigma_3^{\max} = 2\sigma_0,\ \sigma_2^{\max} = \sigma_0.$

Finally, we modify case b) by doubling the density of bar 2, which leads to the problem

$$(\mathbb{SO})_{\rm nf}^{5e} \quad \begin{cases} \min\limits_{x_1,x_2} \ x_1 + x_2 \\ \text{s.t. the constraints in } (\mathbb{SO})_{\rm nf}^{5b}, \end{cases}$$

see Fig. 2.16. The solution point is point B, with the optimal truss lacking bar 2:

$$A_1^* = \frac{F}{2\sigma_0}, \qquad A_2^* = 0,$$

with the optimal weight

$$\frac{FL\rho_0}{\sigma_0}.$$

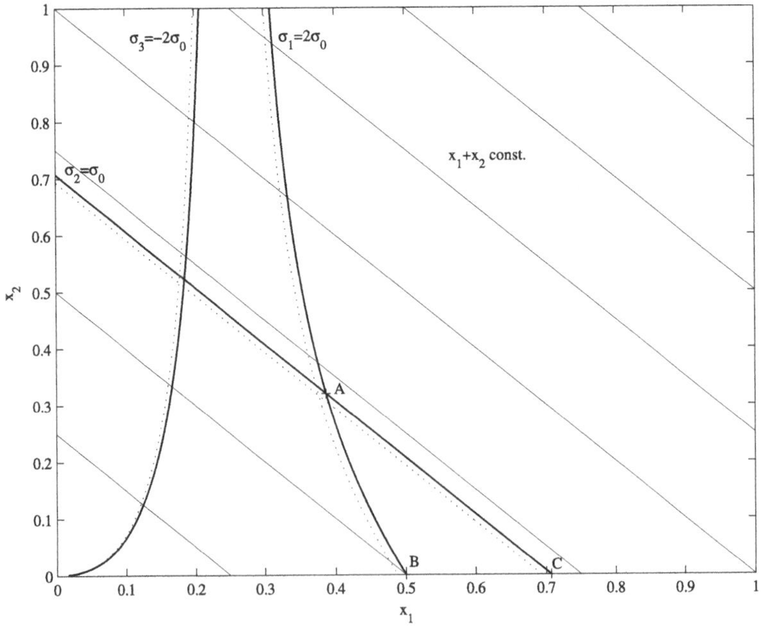

Fig. 2.16 Case e). Point B is the solution

This is the same solution as for case b). The reason that we get the same solution although we have doubled the density of bar 2 is of course that bar 2 is not present in the optimal trusses.

Assume now that A_2 is not allowed to become too small: $A_2 \geq 0.1F/\sigma_0$, i.e. $x_2 \geq 0.1$. Since the σ_2-constraint curve is parallel to the iso-merit lines, we conclude that in this case there will be an infinite number of solutions, namely all points on the line between A and C in Fig. 2.16 for which $x_2 \geq 0.1$! Here, C is the point with $x_1 = 1/\sqrt{2}$ and $x_2 = 0$.

2.6 Weight Minimization of a Three-Bar Truss Subject to a Stiffness Constraint

In this section, the weight of the three-bar truss in the previous section will be minimized under a stiffness constraint; the two-norm of the displacement vector has to be lower than a prescribed value $\delta_0 > 0$, i.e. $\boldsymbol{u}^T\boldsymbol{u} \leq \delta_0^2$. The scalar $\beta = 1/10$, i.e. bar 3 is 10 times longer than bars 1 and 2. The displacements of the free node are given in (2.32)–(2.33). Inserting $\beta = 1/10$ into these expressions we get

$$\boldsymbol{u} = \frac{FL}{EA_1(2A_1 + 11A_2)}\begin{pmatrix} 2A_1 + 10A_2 \\ 10A_2 \end{pmatrix},$$

so that the stiffness constraint may be written as

$$\boldsymbol{u}^T\boldsymbol{u} = \frac{F^2L^2(4A_1^2 + 200A_2^2 + 40A_1A_2)}{E^2A_1^2(2A_1 + 11A_2)^2} \leq \delta_0^2.$$

The density of all bars is ρ_0, which gives the objective function

$$W = \rho_0 L(11A_1 + A_2).$$

Dimensionless variables are introduced according to

$$x_i = \frac{E\delta_0}{FL}A_i, \quad i = 1,\ 2. \tag{2.42}$$

Writing the optimization problem in terms of these variables leads to

$$(\mathbb{SO})_{\text{nf}}^{6} \quad \begin{cases} \min\limits_{x_1,x_2} \ 11x_1 + x_2 \\ \text{s.t.} \quad \begin{cases} \dfrac{4x_1^2 + 200x_2^2 + 40x_1x_2}{x_1^2(2x_1 + 11x_2)^2} - 1 \leq 0 \\ x_1 > 0, \qquad x_2 \geq 0, \end{cases} \end{cases}$$

where we have scaled the objective function by a factor $E\delta_0/(\rho_0 FL^2)$. This problem is illustrated in Fig. 2.17. At first glance it would appear that the solution is

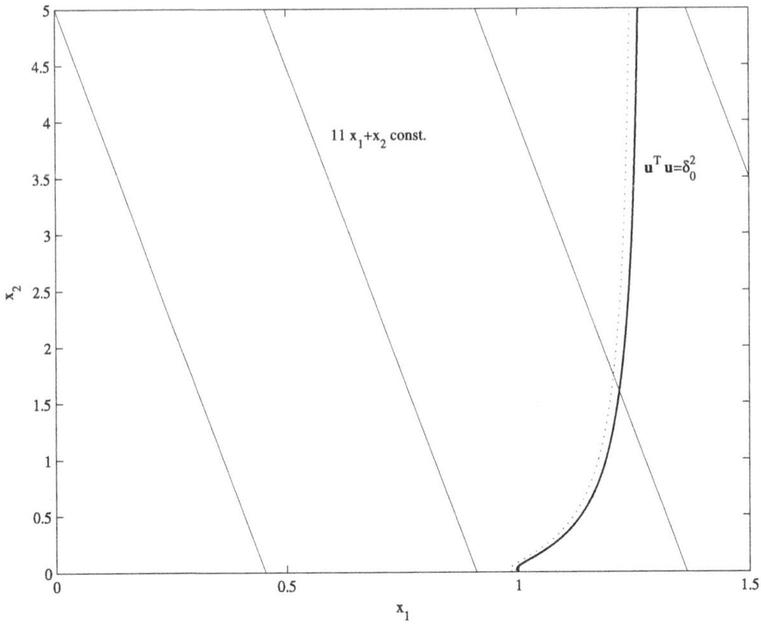

Fig. 2.17 Illustration of problem $(\mathbb{SO})^6_{\mathrm{nf}}$

$x_1 = 1$, $x_2 = 0$. The zoom plots in Fig. 2.18 reveal, however, that this is not the case. The solution may be obtained by first solving the active stiffness constraint equation for x_2 in terms of x_1, and then solving the highly nonlinear one-dimensional optimization problem in the variable x_1 obtained by insertion of the expression for x_2 into the objective function. The solution of the problem is $x_1^* = 0.995$, $x_2^* = 0.0169$. By using (2.42), the corresponding optimal cross-sectional areas are obtained. As a much simpler alternative solution procedure for the two-dimensional optimization problem $(\mathbb{SO})^6_{\mathrm{nf}}$ at hand, we can simply produce finer and finer zoom plots similar to those in Fig. 2.18 and read off the solution.

Since the optimum thickness of bar 2 is very small, it is interesting to investigate how much heavier the optimum structure would be if bar 2 were removed. With no bar 2, the stiffness constraint reads

$$\frac{1}{x_1^2} - 1 \leq 0,$$

whereas the objective function becomes $11x_1$. Thus, with bar 2 removed, the optimal solution is $x_1^* = 1$ and the corresponding (scaled) weight is 11. With bar 2 present, the optimum weight is 10.965, i.e. only 0.3% less than with no bar 2. Since the production cost of the truss would most certainly be significantly less with no bar 2 present, it would make little sense to manufacture the truss with bar 2 included. This serves to illustrate that one should never uncritically accept a solution obtained by performing structural optimization. Finally, we remark that it would have been

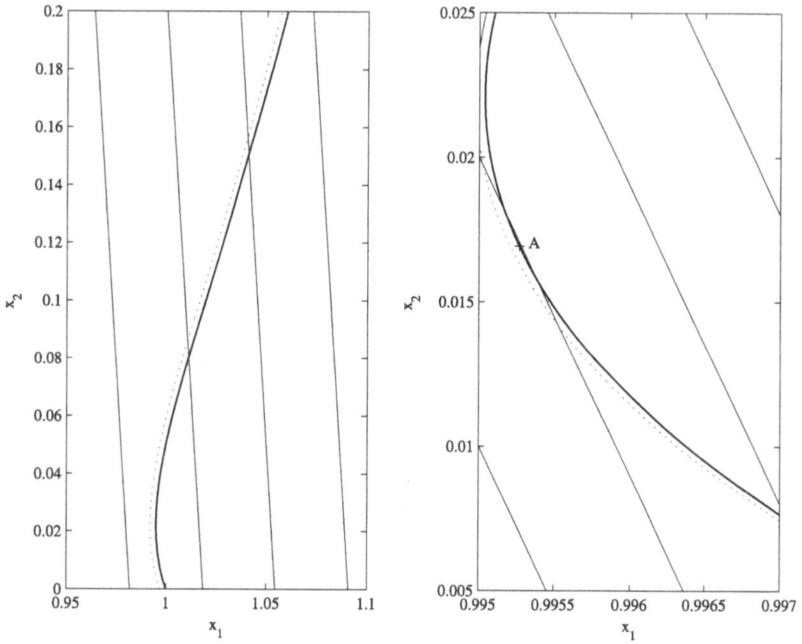

Fig. 2.18 Point A is the solution of problem $(\mathbb{SO})^6_{\text{nf}}$

possible to avoid an optimal solution with a very thin bar 2 if the minimization of the manufacturing cost had, somehow, been included in the optimization problem.

2.7 Exercises

Exercise 2.1 What happens if $F < 0$ in the example of Sect. 2.1?

Exercise 2.2 If the length of the second bar in the example of Sect. 2.5 is changed, the optimum topology of the truss changes: the optimum area of the second bar is zero for $l_2 \geq L$ given $\beta = 1$, $\rho_i = \rho_0$, and $\sigma_i^{\max} = \sigma_0$, $i = 1, 2, 3$. Verify this for a special case, e.g., $l_2 = \sqrt{2}L$.

Exercise 2.3 How does the solution of the example of Sect. 2.5 change if the maximum allowable stress in compression is lower than that in tension?

Exercise 2.4 Verify the details leading to the solutions (x_1^*, x_2^*) and (x_1^{**}, x_2^{**}) in Sect. 2.4.

Exercise 2.5 The stiffness of the two-bar truss subjected to the force $P > 0$ in Fig. 2.19 should be maximized by minimizing the displacement u of the free node.

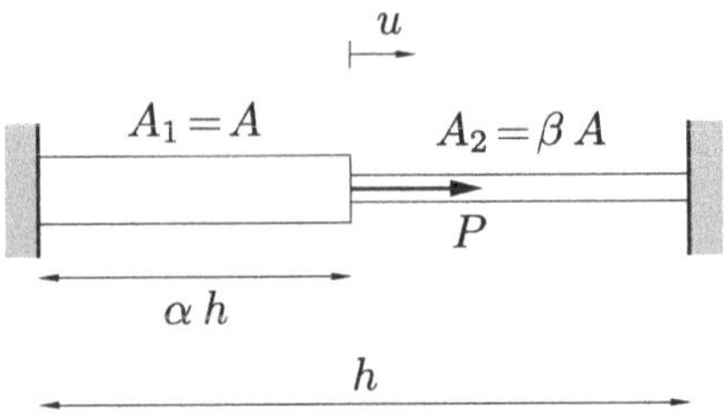

Fig. 2.19 The one-dimensional two-bar truss of Exercise 2.5

Young's modulus is E for both bars. The volume of the truss is not allowed to exceed the value V_0. The total length of the bars is h, and bar 1 has length αh, where α is a scalar between $\alpha_{\min}$ and $\alpha_{\max}$. The cross-sectional areas of the bars are $A_1 = A$ and $A_2 = \beta A$, where β is a scalar. The design variables are α and β. Since α determines the "shape" of the truss, and β the cross-sectional area of bar 2, the problem to be solved is a combined shape and sizing optimization problem.

a) Show that the problem may be formulated as the following mathematical programming problem:

$$\begin{cases} \min\limits_{\alpha,\beta} \dfrac{\alpha(1-\alpha)}{1-\alpha+\alpha\beta} \\ \text{s.t.} \begin{cases} g_1 = \alpha + (1-\alpha)\beta - \dfrac{V_0}{Ah} \leq 0 \\ \alpha_{\min} \leq \alpha \leq \alpha_{\max}, \qquad \beta \geq 0. \end{cases} \end{cases}$$

Let $V_0/(Ah) = 1$ and $\alpha_{\max} = 1$. Show that the set $\{(\alpha, \beta) : g_1 \leq 0, \alpha_{\min} \leq \alpha \leq 1, \beta \geq 0\} = \{(\alpha, \beta) : \alpha_{\min} \leq \alpha \leq 1,\ 0 \leq \beta \leq 1\} \cup \{\alpha, \beta) : \alpha = 1,\ \beta > 1\}$. Solve the problem for arbitrary $\alpha_{\min}$.

b) Let $V_0/(Ah) = 1.2$ and $\alpha_{\min} = 0.2$. Solve the problem for $\alpha_{\max} = 0.6$ and $\alpha_{\max} = 0.8$.

Chapter 3
Basics of Convex Programming

The solution procedure of the previous chapter relies crucially on the ability to easily identify what constraints are active at the solution of the optimization problem under study. This works fine for problems with only two design variables, but when trying to solve real-life problems, where the number of design variables may vary from the order of 10 to the order of 100 000 or more, one needs more systematic solution methods. In this and the following chapter we will study methods from the field of mathematical programming that are applicable for large-scale problems. We begin by reviewing some fundamental results of mathematical programming, with focus on convex programming. Actually, most problems of structural optimization are in fact nonconvex, but this does not imply that convex programming is of little importance in structural optimization: we will see in Chap. 4 that convex approximations play a very important role in the solution algorithms for nonconvex problems. All theorems are presented without proofs; these may be found in any good book on nonlinear mathematical programming such as Bazaraa, Sherali and Shetty [2] or Bertsekas [5].

3.1 Local and Global Optima

Consider a general minimization problem under inequality constraints, where so-called *box constraints* with lower and upper bounds on the variables, are treated separately:

$$(\mathbb{P}) \qquad \begin{cases} \min\limits_{\boldsymbol{x}} g_0(\boldsymbol{x}) \\ \text{s.t.} \quad g_i(\boldsymbol{x}) \leq 0, \qquad i = 1, \ldots, l \\ \qquad\; \boldsymbol{x} \in \mathcal{X}, \end{cases}$$

where $g_i : \mathbb{R}^n \to \mathbb{R}$, $i = 0, \ldots, l$, are assumed to be continuously differentiable functions, and

$$\mathcal{X} = \{\boldsymbol{x} \in \mathbb{R}^n : x_j^{\min} \leq x_j \leq x_j^{\max},\ j = 1, \ldots, n\}.$$

The given lower and upper bounds $x_j^{\min}$ and $x_j^{\max} > x_j^{\min}$ on x_j need not be finite, i.e. the values $x_j^{\min} = -\infty$ and $x_j^{\max} = +\infty$, $j = 1, \ldots, n$, are allowed. Naturally, if all lower and upper bounds are infinite, there are in effect no box constraints. Of course, optimization problems may equally well be written as maximization problems instead. However, any maximization problem may be reformulated as a minimization problem by noting that $\max g_0(\boldsymbol{x}) = -\min(-g_0(\boldsymbol{x}))$. A *feasible point* of ($\mathbb{P}$) is any point, or tuple, $\bar{\boldsymbol{x}}$ in the *feasible set*, i.e. a point that satisfies all the constraints $g_i(\bar{\boldsymbol{x}}) \leq 0,\ i = 1, \ldots, l$ and $\bar{\boldsymbol{x}} \in \mathcal{X}$. Thus, the problem ($\mathbb{P}$) consists of finding a feasible point $\boldsymbol{x}^*$ such that $g_0(\boldsymbol{x}^*) \leq g_0(\bar{\boldsymbol{x}})$ for all feasible points $\bar{\boldsymbol{x}}$ of ($\mathbb{P}$). Such a point is called a *global minimum* of g_0. We note that neither an optimal solution nor any feasible points need exist. For instance, if we were to minimize the tip displacement of the cantilever beam on page 18 without any constraint on the total mass, we would have the following problem:

$$\begin{cases} \min\limits_{x_1, x_2} \ \delta(x_1, x_2) = \dfrac{1}{x_1^3} + \dfrac{7}{x_2^3} \\ \text{s.t.} \quad x_1 > 0, \qquad x_2 > 0. \end{cases} \tag{3.1}$$

For each feasible point $(\bar{x}_1, \bar{x}_2)$ we can find another feasible point $(\bar{\bar{x}}_1, \bar{\bar{x}}_2)$ with $\bar{\bar{x}}_1 > \bar{x}_1$ and $\bar{\bar{x}}_2 > \bar{x}_2$ such that $\delta(\bar{\bar{x}}_1, \bar{\bar{x}}_2) < \delta(\bar{x}_1, \bar{x}_2)$, and consequently no minimum exists. Further, if we want to minimize the tip displacement under a mass constraint, and x_1 and x_2 have to be greater than some specified values $x_1^{\min}$ and $x_2^{\min}$, respectively, we would have the problem

$$\begin{cases} \min\limits_{x_1, x_2} \ \dfrac{1}{x_1^3} + \dfrac{7}{x_2^3} \\ \text{s.t.} \quad \begin{cases} C_1(x_1 + x_2) \leq W \\ x_1 \geq x_1^{\min} > 0, \qquad x_2 \geq x_2^{\min} > 0. \end{cases} \end{cases} \tag{3.2}$$

Then, if $x_1^{\min} + x_2^{\min} > W/C_1$, no feasible point exists. Naturally, an optimum cannot exist when there are no feasible points.

In general it is extremely computationally demanding to determine a global minimum. Instead, we will rest content with trying to obtain a *local minimum*. A point $\boldsymbol{x}^*$ is a local minimum if the objective function g_0 only assumes greater or equal values in a surrounding of $\boldsymbol{x}^*$, but may very well assume smaller values elsewhere. Naturally, any global minimum is also a local minimum. For unconstrained optimization problems, local (and global) minima are located at stationary points $\boldsymbol{x}^*$, i.e. points for which the gradient of g_0 is zero:

$$\nabla g_0(\boldsymbol{x}^*) = \begin{bmatrix} \dfrac{\partial g_0(\boldsymbol{x}^*)}{\partial x_1} \\ \vdots \\ \dfrac{\partial g_0(\boldsymbol{x}^*)}{\partial x_n} \end{bmatrix} = \boldsymbol{0}.$$

A stationary point need not be a local minimum, however; it may equally well be a local maximum. For constrained problems, local minima are not even necessarily located at stationary points, as they may be located on the boundary of the feasible set. In order to illustrate the difference between stationary points, local and global optima, we consider a function $g_0 : [x_1,\ x_6] \rightarrow \mathbb{R}$, cf. Fig. 3.1.[1] The points x_2,

[1] Here, $[a,\ b]$ denotes the interval $\{x : a \leq x \leq b\}$. Similarly, $(a,\ b) = \{x : a < x < b\}$ and $[a,\ b) = \{x : a \leq x < b\}$.

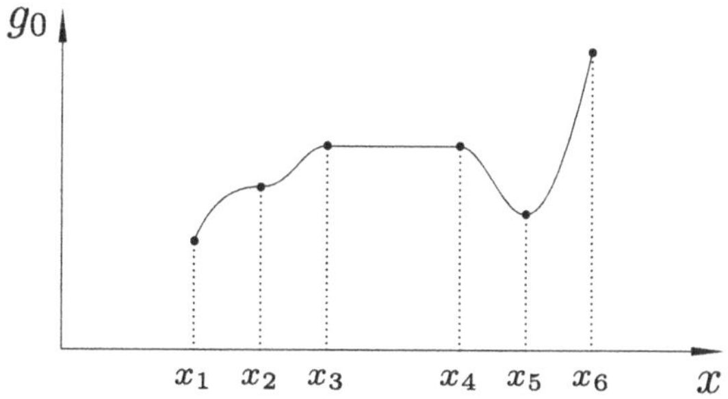

Fig. 3.1 A function with several local optima

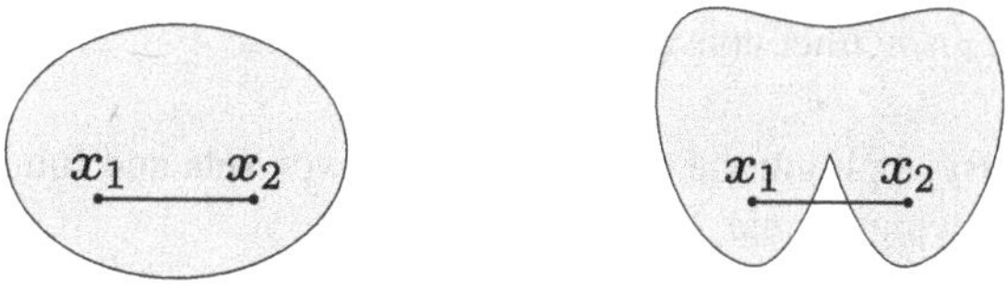

Fig. 3.2 A convex (*left*) and a nonconvex (*right*) set

$[x_3, x_4]$ and x_5 are stationary points, x_1, (x_3, x_4) and x_5 are local minima, $[x_3, x_4]$ and x_6 are local maxima. Point x_2 is neither a local minimum nor a local maximum. Point x_1 is the global minimum and x_6 is the global maximum.

Let us return to the three-bar truss on page 21 and see if there are any local optima that are not global optima for the four different cases studied. In Fig. 2.12, point A is the unique global minimum. In Fig. 2.13, B is the unique global minimum and A is a local minimum. In Fig. 2.14, both A and B are global minima. In Fig. 2.15, A is the unique global minimum and B is a local minimum. Finally, in Fig. 2.16, B is the unique global minimum, and all points on the line from A to C, not including the end point C itself, are local minima with identical objective function values.

Although for general problems local minima are not necessarily global minima, there is an important class of problems for which they are: convex problems.

3.2 Convexity

A set $\mathcal{S} \subset \mathbb{R}^n$ is *convex* if for all $\boldsymbol{x}_1, \boldsymbol{x}_2 \in S$ and all $\lambda \in (0, 1)$, it holds that

$$\lambda \boldsymbol{x}_1 + (1 - \lambda)\boldsymbol{x}_2 \in \mathcal{S}.$$

Thus, a set is convex if all points on the line connecting any two points in the set also belongs to the set, cf. Fig. 3.2. A function $f : S \to \mathbb{R}$ is *convex* (on the convex set $\mathcal{S} \subset \mathbb{R}^n$) if for all $\boldsymbol{x}_1, \boldsymbol{x}_2 \in \mathcal{S}$ with $\boldsymbol{x}_1 \neq \boldsymbol{x}_2$ and all $\lambda \in (0, 1)$, it holds that

$$f(\lambda \boldsymbol{x}_1 + (1 - \lambda)\boldsymbol{x}_2) \leq \lambda f(\boldsymbol{x}_1) + (1 - \lambda) f(\boldsymbol{x}_2).$$

Similarly, f is *strictly convex* if strict inequality ($<$) holds above instead. f is (strictly) *concave* if $-f$ is (strictly) convex. The graph of a convex function thus lies on or below the straight line connecting any two points on this graph. For a strictly convex function, the graph lies strictly below the line, cf. Fig. 3.3.

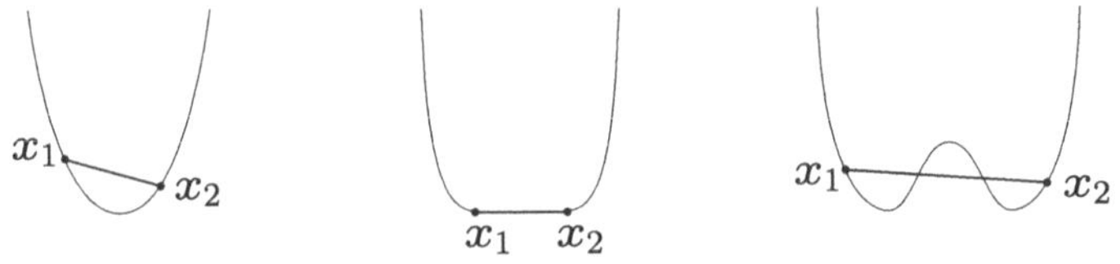

Fig. 3.3 A strictly convex (*left*), a convex (*middle*), and a nonconvex (*right*) function

Example 3.1 The function $f : \mathbb{R} \to \mathbb{R},\ f(x_1) = x_1^2$ is strictly convex, but $f : \mathbb{R}^2 \to \mathbb{R},\ f(x_1, x_2) = x_1^2$ is only convex. The function $f : \mathbb{R}^2 \to \mathbb{R},\ f(x_1, x_2) = x_1 x_2$ is neither convex nor concave.

By applying the definitions of convex sets and functions, one easily obtains the following lemma.

Lemma 3.1 (i) *The set* $\mathcal{S} = \{\boldsymbol{x} \in \mathcal{X} : g_i(\boldsymbol{x}) \le 0,\ i = 1, \dots, l\}$ *is convex if the functions* $g_i : \mathbb{R}^n \to \mathbb{R},\ i = 1, \dots, l$ *are convex.*

(ii) *Let* $\mathcal{S}$ *be a convex set. If* $f : \mathcal{S} \to \mathbb{R}$ *and* $g : \mathcal{S} \to \mathbb{R}$ *are convex and* $h : \mathcal{S} \to \mathbb{R}$ *is strictly convex, then* αf *is convex, where* $\alpha \ge 0$ *is an arbitrary scalar,* $f + g$ *is convex and* $f + h$ *is strictly convex.*

If both the objective function and the feasible set of ($\mathbb{P}$) are convex, the problem is said to be convex. The lemma above then states that ($\mathbb{P}$) is convex if the objective function and all constraint functions $g_i,\ i = 1, \dots, l$, are convex.

As previously mentioned, local minima are also global minima for convex problems. However, as indicated by the convex problems in (3.1) and (3.2), convex problems need not have a solution (you are to demonstrate the convexity of these problems in Exercise 3.1). When the feasible set is compact, i.e. bounded and closed, a solution always exists (this is true for any continuous objective function, not necessarily convex). If the objective function is strictly convex and the feasible set is convex, there exists at most one solution. If, in addition, the feasible set is compact, there exists exactly one solution. For example, if the strictly convex function $1/x$ is minimized over the closed, but unbounded convex set $x \ge 1$, no solution exists. If the same function is minimized over the compact set $[1,\ 2]$, the solution is $x^* = 1/2$. Note that the convexity of the feasible set plays a crucial role here; for example if the strictly convex function $x_1^2 + x_2^2$ is minimized over the nonconvex, compact set $1 \le x_1^2 + x_2^2 \le 2$, there is an infinite number of global minima, namely all points (x_1^*, x_2^*) with $(x_1^*)^2 + (x_2^*)^2 = 1$.

In order to determine whether a continuously differentiable function is convex, we may study its gradient.

Theorem 3.1 *Let* $f : \mathcal{S} \to \mathbb{R}$, *where* $\mathcal{S}$ *is convex and* f *is continuously differentiable. Then* f *is (strictly) convex if and only if the gradient* ∇f *is (strictly) monotone.*

Here, a function $\boldsymbol{g} : \mathcal{S} \to \mathbb{R}^n$ is *monotone* on $\mathcal{S}$ if for all $\boldsymbol{x}_1, \boldsymbol{x}_2 \in \mathcal{S}$ with $\boldsymbol{x}_1 \neq \boldsymbol{x}_2$ it holds that

$$(\boldsymbol{x}_2 - \boldsymbol{x}_1)^T (\boldsymbol{g}(\boldsymbol{x}_2) - \boldsymbol{g}(\boldsymbol{x}_1)) \geq 0.$$

Similarly, $\boldsymbol{g}$ is *strictly monotone* on $\mathcal{S}$ if strict inequality holds here. This definition is a generalization of the concept of a monotonically increasing function of one variable: g is monotonically increasing if $x_2 > x_1$ implies that $g(x_2) \geq g(x_1)$.

Example 3.2 The function $f : \mathbb{R} \to \mathbb{R}$, $f(x) = x^4$, is strictly convex on $\mathbb{R}$ since $\nabla f(x) = 4x^3$ is strictly monotone on $\mathbb{R}$:

$$\begin{aligned}(x_2 - x_1)(x_2^3 - x_1^3) &= (x_2 - x_1)^2(x_1^2 + x_1 x_2 + x_2^2) \\ &= (x_2 - x_1)^2\left[\left(x_1 + \frac{1}{2}x_2\right)^2 + \frac{3}{4}x_2^2\right] > 0, \quad x_1 \neq x_2.\end{aligned}$$

For a twice differentiable function, convexity is most efficiently tested by examining its Hessian.

Theorem 3.2 *Let $f : \mathcal{S} \to \mathbb{R}$, where $\mathcal{S}$ is convex and f is twice continuously differentiable. Then*

(i) *f is convex if and only if the Hessian $\nabla^2 f$ is positive semidefinite,*
(ii) *f is strictly convex if $\nabla^2 f$ is positive definite.*

Here, the Hessian is given by

$$\nabla^2 f(\boldsymbol{x}) = \begin{bmatrix} \frac{\partial^2 f(\boldsymbol{x})}{\partial x_1^2} & \frac{\partial^2 f(\boldsymbol{x})}{\partial x_1 \partial x_2} & \cdots & \frac{\partial^2 f(\boldsymbol{x})}{\partial x_1 \partial x_n} \\ \frac{\partial^2 f(\boldsymbol{x})}{\partial x_2 \partial x_1} & \frac{\partial^2 f(\boldsymbol{x})}{\partial x_2^2} & \cdots & \frac{\partial^2 f(\boldsymbol{x})}{\partial x_2 \partial x_n} \\ \vdots & \vdots & \ddots & \vdots \\ \frac{\partial^2 f(\boldsymbol{x})}{\partial x_n \partial x_1} & \frac{\partial^2 f(\boldsymbol{x})}{\partial x_n \partial x_2} & \cdots & \frac{\partial^2 f(\boldsymbol{x})}{\partial x_n^2} \end{bmatrix},$$

a matrix $\boldsymbol{A} \in \mathbb{R}^{n \times n}$ is positive semidefinite if

$$\boldsymbol{y}^T \boldsymbol{A} \boldsymbol{y} \geq 0,$$

for all $\boldsymbol{y} \in \mathbb{R}^n$, and positive definite if

$$\boldsymbol{y}^T \boldsymbol{A} \boldsymbol{y} > 0,$$

for all $\boldsymbol{y} \in \mathbb{R}^n$ with $\boldsymbol{y} \neq \boldsymbol{0}$. The positive definiteness of a symmetric matrix may be checked using Sylvester's criterion:

Theorem 3.3 *A symmetric matrix $\boldsymbol{A} \in \mathbb{R}^{n \times n}$ is positive definite if and only if the determinant of the upper left $k \times k$ submatrix is positive for each $k = 1, \ldots, n$.*

Alternatively, one may use the fact that a symmetric matrix is positive definite if and only if it possesses a Cholesky decomposition, i.e. that it may be written as $\boldsymbol{A} = \boldsymbol{L}\boldsymbol{L}^T$, where $\boldsymbol{L}$ is a nonsingular lower triangular matrix.

Example 3.3 Consider the function $f : \mathbb{R}^2 \to \mathbb{R}$, $f(x_1, x_2) = x_1^2 + x_2^2$. Then

$$\nabla f(x_1, x_2) = \begin{bmatrix} 2x_1 \\ 2x_2 \end{bmatrix}, \qquad \nabla^2 f(x_1, x_2) = \begin{bmatrix} 2 & 0 \\ 0 & 2 \end{bmatrix}.$$

The Hessian is positive definite by Sylvester's criterion since $2 > 0$ and $2 \cdot 2 - 0 \cdot 0 > 0$. Thus, Theorem 3.2(ii) implies that f is strictly convex on $\mathbb{R}^2$.

Example 3.4 The function $f : \mathbb{R} \to \mathbb{R}$, $f(x) = x^4$ has the Hessian $\nabla^2 f(x) = 12x^2 > 0$, $x \neq 0$. Thus, from Theorem 3.2(i), we conclude that f is (at least) convex. At the origin, the Hessian is zero, and consequently not positive definite, so from Theorem 3.2(ii), we cannot conclude that f is strictly convex. We know, however, from analyzing the gradient ∇f in Example 3.2, that f *is* indeed strictly convex.

We end this section by investigating which of the problems studied in Chap. 2 are convex. In problem $(\mathbb{SO})^1_{\text{nf}}$ on page 11, where the weight of a statically determinate two-bar truss was minimized subject to stress constraints, the objective function is a linear, i.e. also a convex, function, and the constraints are box constraints which are always convex. Thus, $(\mathbb{SO})^1_{\text{nf}}$ is a convex problem.

We leave it as an exercise to show that the weight minimization problem $(\mathbb{SO})^2_{\text{nf}}$ of a statically determinate two-bar truss under stress and instability constraints represents a convex problem, as is the weight minimization of a statically determinate two-bar truss under stress and displacement constraints in $(\mathbb{SO})^3_{\text{nf}}$ on page 17. The problem $(\mathbb{SO})^4_{\text{nf}}$ on page 20 where the weight of a two-beam cantilever was minimized subject to a displacement constraint is also convex.

In the next problem, the weight of a statically indeterminate three-bar truss was minimized under stress constraints. In this case the stress constraints could not be written as simple box constraints. It turns out that the convexity properties of this problem depends on the relation between the stress limits of the bars. For problem $(\mathbb{SO})^{5a}_{\text{nf}}$ on page 25, the only constraints that limit the feasible set are the σ_1-constraint and $x_2 \geq 0$. The σ_1-constraint is written $g_1(x_1, x_2) \leq 0$ where

$$g_1(x_1, x_2) = \frac{2x_1 + x_2}{2x_1(x_1 + x_2)} - 1.$$

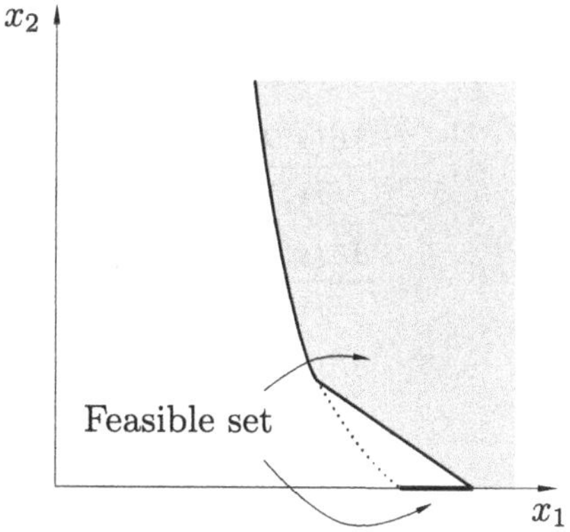

Fig. 3.4 The feasible set for problems $(\mathbb{SO})^{5b}_{\text{nf}}$–$(\mathbb{SO})^{5e}_{\text{nf}}$

A straightforward calculation yields the Hessian as

$$\nabla^2 g_1(x_1, x_2) = \frac{1}{(x_1 + x_2)^3} \begin{bmatrix} \dfrac{(2x_1 + x_2)(x_1^2 + x_1 x_2 + x_2^2)}{x_1^3} & 1 \\ 1 & 1 \end{bmatrix}.$$

Thus, the (1, 1)-component of the Hessian is positive in the feasible set, and, in addition, the determinant of the Hessian is $x_1^{-3}(x_1 + x_2)^{-3}$ which is also positive in the feasible set. By Sylvester's criterion, Theorem 3.3, the Hessian is positive definite, and Theorem 3.2 then states that g_1 is strictly convex. Since the objective function is also linear, we conclude that problem $(\mathbb{SO})^{5a}_{\text{nf}}$ is convex. Note that the σ_3-constraint is not convex, but concave. This, however, has no influence on the convexity of the problem as the σ_3-constraint does not limit the feasible set.

The problems $(\mathbb{SO})^{5b}_{\text{nf}}$–$(\mathbb{SO})^{5e}_{\text{nf}}$ are all nonconvex, since, as indicated in Fig. 3.4, the feasible sets are not convex. It is the nonconvexity of these problems that gives rise to local optima that are not global optima. We note that if one introduces a positive lower bound $x_2^{\min}$, then these problem are all convex as the "tail" that is responsible for the nonconvexity in that case is not part of the feasible set.

The weight minimization $(\mathbb{SO})^{6}_{\text{nf}}$ of a statically indeterminate three-bar truss under a constraint of the two-norm of the displacement vector is clearly a nonconvex problem as is evident from Fig. 2.17.

3.3 KKT Conditions

We now turn to the question of how to identify a local, i.e. also a global, minimum of a convex optimization problem. To that end we first define the *Lagrangian function* $\mathcal{L} : \mathbb{R}^n \times \mathbb{R}^l \to \mathbb{R}$ of $(\mathbb{P})$ on page 35 as

$$\mathcal{L}(\boldsymbol{x}, \boldsymbol{\lambda}) = g_0(\boldsymbol{x}) + \sum_{i=1}^{l} \lambda_i g_i(\boldsymbol{x}), \tag{3.3}$$

where $\lambda_i,\ i = 1, \ldots, l$ are called Lagrange multipliers. The *Karush–Kuhn–Tucker (KKT) conditions* of $(\mathbb{P})$ are defined as

$$\frac{\partial \mathcal{L}(\boldsymbol{x}, \boldsymbol{\lambda})}{\partial x_j} \leq 0, \quad \text{if } x_j = x_j^{\max}, \tag{3.4}$$

$$\frac{\partial \mathcal{L}(\boldsymbol{x}, \boldsymbol{\lambda})}{\partial x_j} = 0, \quad \text{if } x_j^{\min} < x_j < x_j^{\max}, \tag{3.5}$$

$$\frac{\partial \mathcal{L}(\boldsymbol{x}, \boldsymbol{\lambda})}{\partial x_j} \geq 0, \quad \text{if } x_j = x_j^{\min}, \tag{3.6}$$

$$\lambda_i g_i(\boldsymbol{x}) = 0, \tag{3.7}$$

$$g_i(\boldsymbol{x}) \leq 0, \tag{3.8}$$

$$\lambda_i \geq 0, \tag{3.9}$$

$$\boldsymbol{x} \in \mathcal{X}, \tag{3.10}$$

for all $j = 1, \ldots, n$ and $i = 1, \ldots, l$. Partial differentiation of $\mathcal{L}$ with respect to the design variables gives

$$\frac{\partial \mathcal{L}(\boldsymbol{x}, \boldsymbol{\lambda})}{\partial x_j} = \frac{\partial g_0(\boldsymbol{x})}{\partial x_j} + \sum_{i=1}^{l} \lambda_i \frac{\partial g_i(\boldsymbol{x})}{\partial x_j}.$$

In most texts, box constraints are not treated separately, but are instead included in $g_i(\boldsymbol{x}) \leq 0,\ i = 1, \ldots, l$, by writing $x_j - x_j^{\max} \leq 0$ and $x_j^{\min} - x_j \leq 0,\ j = 1, \ldots, n$. The Lagrangian multipliers corresponding to these constraints may easily be eliminated, however, leading to the KKT conditions above. From (3.7) it is seen that if a constraint g_i is not active, i.e. $g_i(\boldsymbol{x}) \neq 0$, then the corresponding $\lambda_i = 0$. Similarly, if $\lambda_i \neq 0$, then g_i is active: $g_i(\boldsymbol{x}) = 0$. Each point $(\boldsymbol{x}^*, \boldsymbol{\lambda}^*) \in \mathbb{R}^n \times \mathbb{R}^l$ satisfying (3.4)–(3.10) is called a KKT point. For sufficiently regular nonconvex problems, the KKT conditions are necessary, but not sufficient, optimality conditions for $(\mathbb{P})$. That is, local optima are always found among the KKT points, but there may be KKT points that are not local optima. The fact that the KKT conditions cannot be sufficient for optimality is evident by studying the special case of an unconstrained optimization problem, where the KKT points are equivalent to stationary points. Numerical algorithms typically try to find KKT points, and thus one may end up at a point that is not a local minimum, but even a local maximum! However, for convex problems a KKT point is always an optimal point. One has the following theorems.

Theorem 3.4 *Let the problem* $(\mathbb{P})$ *be convex and satisfy Slater's constraint qualification (CQ), i.e. there exists a point* $\hat{\boldsymbol{x}} \in \mathcal{X}$ *such that* $g_i(\hat{\boldsymbol{x}}) < 0, i = 1, \ldots, l$. *Let* $\boldsymbol{x}^*$ *be a local (i.e. also global) minimum of* $(\mathbb{P})$. *Then there exists a* $\boldsymbol{\lambda}^*$ *such that* $(\boldsymbol{x}^*, \boldsymbol{\lambda}^*)$ *is a KKT point of* $(\mathbb{P})$.

Theorem 3.5 *Let* $(\mathbb{P})$ *be a convex problem, and let* $(\boldsymbol{x}^*, \boldsymbol{\lambda}^*)$ *be a KKT point of* $(\mathbb{P})$. *Then* $\boldsymbol{x}^*$ *is a local (i.e. also global) minimum of* $(\mathbb{P})$.

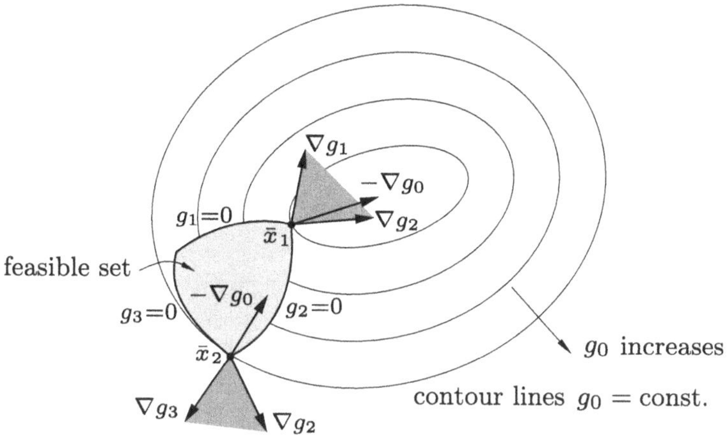

Fig. 3.5 Illustration of the KKT conditions

A geometric interpretation of this theorem when there are no box constraints is given in Fig. 3.5. The KKT conditions state that $-\nabla g_0(\bar{x})$ should belong to the cone spanned[2] by the gradients of the active constraints at a point $\bar{x}$. At point $\bar{x}_1$ in the figure, $-\nabla g_0(\bar{x}_1) = \lambda_1 \nabla g_1(\bar{x}_1) + \lambda_2 \nabla g_2(\bar{x}_1)$, $\lambda_1 \geq 0$, $\lambda_2 \geq 0$, so $\bar{x}_1$ is a KKT point and consequently the optimal solution. Regarding point $\bar{x}_2$, we see that $-\nabla g_0(\bar{x}_2)$ does not belong to the cone spanned by the active constraints at $\bar{x}_2$. That is, there does not exist any $\lambda_2 \geq 0$ and $\lambda_3 \geq 0$ such that $-\nabla g_0(\bar{x}_2) = \lambda_2 \nabla g_2(\bar{x}_2) + \lambda_3 \nabla g_3(\bar{x}_2)$, and consequently $\bar{x}_2$ is not a KKT point. Theorem 3.4 then implies that $\bar{x}_2$ is not an optimum.

Example 3.5 We study the two-bar truss on page 14 again, but this time we will solve the optimization problem by calculating a KKT point instead. It has been shown that the problem of minimizing the weight of the truss subject to displacement and stress constraints may be written as

$$\min_{A_1,A_2} \rho L \left(\frac{2}{\sqrt{3}} A_1 + A_2 \right)$$

$$\text{s.t.} \quad \begin{cases} F\left(\dfrac{8}{\sqrt{3}A_1} + \dfrac{3}{A_2} \right) \leq \sigma_0 \\ -\sigma_0 \leq \dfrac{2F}{A_1} \leq \sigma_0 \\ -\sigma_0 \leq -\dfrac{\sqrt{3}F}{A_2} \leq \sigma_0 \\ A_1 \geq 0, \; A_2 \geq 0. \end{cases}$$

---

[2]The cone spanned by some vectors $v_1, \ldots, v_l$ is the set of nonnegative linear combinations of these vectors.

We rewrite the problem in the same form as the general problem $(\mathbb{P})$:

$$
(\mathbb{P})_1 \begin{cases} \min\limits_{A_1,A_2} g_0 = \dfrac{2}{\sqrt{3}}A_1 + A_2 \\ \text{s.t.} \begin{cases} g_1 = F\left(\dfrac{8}{\sqrt{3}A_1} + \dfrac{3}{A_2}\right) - \sigma_0 \le 0 \\ g_2 = \dfrac{2F}{A_1} - \sigma_0 \le 0 \\ g_3 = -\dfrac{2F}{A_1} - \sigma_0 \le 0 \\ g_4 = -\dfrac{\sqrt{3}F}{A_2} - \sigma_0 \le 0 \\ g_5 = \dfrac{\sqrt{3}F}{A_2} - \sigma_0 \le 0 \\ (A_1, A_2) \in \mathcal{X} = \{(A_1, A_2) \,:\, A_1 \ge 0,\ A_2 \ge 0\}, \end{cases} \end{cases}
$$

where we, for convenience, have skipped the arguments in $g_i(A_1, A_2)$, $i = 1, \dots, 5$. Also, the term ρL has been dropped from the objective function since it is only a positive constant, and, hence, does not affect the optimal solution (A_1^*, A_2^*). The g_1-constraint is the displacement constraint, g_2 and g_3 are the σ_1-constraints, and g_4 and g_5 are the σ_2-constraints. Next, the KKT conditions (3.4)–(3.10) will be formulated. We conclude immediately that $A_1 \neq 0$ at the solution, otherwise the constraint $2F/A_1 \le \sigma_0$ would be violated. Similarly, $A_2 \neq 0$ at the solution, so that the constraint $-\sigma_0 \le -\sqrt{3}F/A_2$ is not violated. Since $F > 0$, $\sigma_0 > 0$, $A_1 > 0$, $A_2 > 0$, the constraints g_3 and g_4 can never be active, i.e. it always holds that $g_3 < 0$ and $g_4 < 0$. Consequently, the corresponding Lagrangian multipliers, λ_3 and λ_4 are both zero. We also see that

$$
g_2 = \frac{2F}{A_1} - \sigma_0 < F\left(\frac{8}{\sqrt{3}A_1} + \frac{3}{A_2}\right) - \sigma_0 = g_1 \le 0,
$$

so g_2 can never be active either: $\lambda_2 = 0$. Similarly,

$$
g_5 = \frac{\sqrt{3}F}{A_2} - \sigma_0 < F\left(\frac{8}{\sqrt{3}A_1} + \frac{3}{A_2}\right) - \sigma_0 = g_1 \le 0,
$$

so $\lambda_5 = 0$ as well. Thus, only λ_1 may be nonzero. Since the cross-sectional areas cannot be zero at the solution, the KKT conditions become

$$
\begin{bmatrix} \dfrac{2}{\sqrt{3}} \\ 1 \end{bmatrix} + \lambda_1 \begin{bmatrix} -\dfrac{8}{\sqrt{3}A_1^2} \\ -\dfrac{3}{\sqrt{3}A_2^2} \end{bmatrix} = \begin{bmatrix} 0 \\ 0 \end{bmatrix}, \tag{3.11}
$$

$$
\lambda_1 \left(\frac{8F}{\sqrt{3}A_1} + \frac{3F}{A_2} - \sigma_0\right) = 0. \tag{3.12}
$$

From the second row in (3.11) we get

$$\lambda_1 = \frac{A_2^2}{3} \neq 0.$$

Insertion of this into the first row gives

$$\frac{2}{\sqrt{3}} - \frac{8A_2^2}{3\sqrt{3}A_1^2} = 0,$$

from which we get

$$A_2 = \frac{\sqrt{3}}{2} A_1.$$

If we insert this into (3.12), we finally get

$$A_1^* = \frac{14F}{\sqrt{3}\sigma_0} \quad \text{and} \quad A_2^* = \frac{7F}{\sigma_0}.$$

Since the problem is convex, cf. Exercise 3.1, we know that this KKT point is also the global minimum of $(\mathbb{P})_1$, see Fig. 3.6.

In $(\mathbb{P})_1$ we have treated the stress constraints as general constraints, i.e. they were written as $g_2 \leq 0, \ldots, g_5 \leq 0$. However, since in this example (and as in any statically determinate truss), the stress in each bar depends only on the cross-sectional area of that particular bar, the stress constraints may be written as simple box constraints instead, cf. (2.17). If we take advantage of this fact, the following optimization problem is obtained immediately:

$$(\mathbb{P})_2 \quad \begin{cases} \min\limits_{A_1, A_2} g_0 = \dfrac{2}{\sqrt{3}} A_1 + A_2 \\ \text{s.t.} \begin{cases} g_1 = F\left(\dfrac{8}{\sqrt{3}A_1} + \dfrac{3}{A_2}\right) - \sigma_0 \leq 0 \\ (A_1, A_2) \in \mathcal{X} = \left\{ (A_1, A_2) \, : \, A_1 \geq \dfrac{2F}{\sigma_0}, \; A_2 \geq \dfrac{\sqrt{3}F}{\sigma_0} \right\}. \end{cases} \end{cases}$$

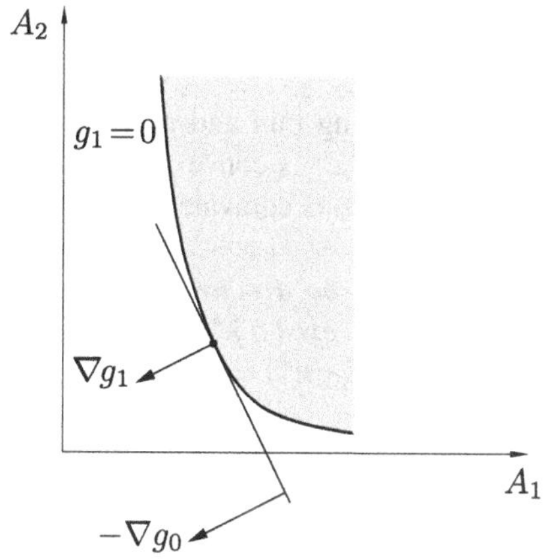

Fig. 3.6 Geometric illustration of $(\mathbb{P})_1$. At the solution, $-\nabla g_0 = \lambda_1 \nabla g_1$, $\lambda_1 > 0$

The KKT conditions of $(\mathbb{P})_2$ are identical to (3.11) and (3.12) since the box constraints cannot be active.

3.4 Lagrangian Duality

It may be quite tedious to obtain an optimal solution of $(\mathbb{P})$ by solving the nonlinear equations and inequalities constituting the KKT conditions (3.4)–(3.10) directly. In this section, we will therefore describe another method to obtain an optimal solution that will prove more suitable, especially for large-scale structural optimization problems.

It may be proven that $(\mathbb{P})$ is equivalent to the following min-max problem:

$$(\mathbb{P}_L) \quad \min_{x \in \mathcal{X}} \max_{\boldsymbol{\lambda} \geq \mathbf{0}} \mathcal{L}(\boldsymbol{x}, \boldsymbol{\lambda}) = \min_{x \in \mathcal{X}} \max_{\boldsymbol{\lambda} \geq \mathbf{0}} \left\{ g_0(\boldsymbol{x}) + \sum_{i=1}^{l} \lambda_i g_i(\boldsymbol{x}) \right\}.$$

Thus, first the Lagrangian $\mathcal{L}$ of $(\mathbb{P})$ is maximized with respect to $\boldsymbol{\lambda} \geq \mathbf{0}$ for a fixed $\boldsymbol{x}$, and the result is then minimized with respect to $\boldsymbol{x} \in \mathcal{X}$. Note that the result of the maximization will be $+\infty$ if some $g_i(\boldsymbol{x}) > 0$, and $g_0(\boldsymbol{x})$ if all $g_i(\boldsymbol{x}) \leq 0$, $i = 1, \ldots, l$. Therefore, when the result of the maximization is minimized with respect to $\boldsymbol{x} \in \mathcal{X}$, the solution of $(\mathbb{P})$ is obtained. Nothing much has been gained by this reformulation, however. As will be described in the next chapter, we will approximate the objective functions and constraints of our structural optimization problems in such a way that it is computationally efficient to solve the so-called (Lagrangian) dual problem $(\mathbb{D})$ corresponding to the primal problem $(\mathbb{P})$, which is obtained by interchanging min and max in $(\mathbb{P}_L)$:

$$(\mathbb{D}) \quad \begin{cases} \max\limits_{\boldsymbol{\lambda}} \varphi(\boldsymbol{\lambda}) \\ \text{s.t.} \quad \boldsymbol{\lambda} \geq \mathbf{0}, \end{cases}$$

where the dual objective function φ is defined as

$$\varphi(\boldsymbol{\lambda}) = \min_{x \in \mathcal{X}} \mathcal{L}(\boldsymbol{x}, \boldsymbol{\lambda}).$$

In general, interchanging min and max in $(\mathbb{P}_L)$, results in a completely different problem. However, if $(\mathbb{P})$ is convex and Slater's CQ is satisfied, cf. Theorem 3.4, then it turns out that $(\mathbb{D})$ is equivalent to $(\mathbb{P}_L)$, and, thus, to $(\mathbb{P})$:

Theorem 3.6 *Let* $(\mathbb{P})$ *be a convex problem with the set* $\mathcal{X}$ *compact, satisfying Slater's CQ. Then there exist a* $\boldsymbol{\lambda}^*$ *that solves* $(\mathbb{D})$, *and an* $\boldsymbol{x}^* \in \operatorname{argmin}_{x \in \mathcal{X}} \mathcal{L}(\boldsymbol{x}, \boldsymbol{\lambda}^*)$ *that solves* $(\mathbb{P})$, *where* $g_0(\boldsymbol{x}^*) = \varphi(\boldsymbol{\lambda}^*)$.

In order to solve $(\mathbb{P})$ we may thus solve $(\mathbb{D})$ instead, i.e. solve a min–max problem. It should be noted that the constraints in these optimizations are very simple: $\boldsymbol{x} \in \mathcal{X}$ and $\boldsymbol{\lambda} \geq \mathbf{0}$, respectively. This is a major advantage of duality theory, since in $(\mathbb{P})$ we have the constraints $g_i(\boldsymbol{x}) \leq 0,\ i = 1, \ldots, l$, that may be very complicated to deal with directly. The problem of maximizing φ is not only easy because of the simple constraints, but also because φ is always concave. If the problem $\min_{\boldsymbol{x}\in\mathcal{X}} \mathcal{L}(\boldsymbol{x}, \boldsymbol{\lambda})$ has exactly one solution for a given $\boldsymbol{\lambda}$ (a sufficient condition for this is that g_0 is strictly convex and $\mathcal{X}$ is compact), then φ is differentiable at $\boldsymbol{\lambda}$, and it holds that

$$\frac{\partial \varphi(\boldsymbol{\lambda})}{\partial \lambda_i} = g_i(\boldsymbol{x}^*(\boldsymbol{\lambda})), \quad i = 1, \ldots, l, \text{ where } \boldsymbol{x}^*(\boldsymbol{\lambda}) = \min_{\boldsymbol{x}\in\mathcal{X}} \mathcal{L}(\boldsymbol{x}, \boldsymbol{\lambda}), \tag{3.13}$$

which is a useful property when maximizing φ.

3.4.1 Lagrangian Duality for Convex and Separable Problems

We consider an optimization problem of the following form

$$(\mathbb{P})_s \qquad \begin{cases} \min\limits_{\boldsymbol{x}} g_0(\boldsymbol{x}) \\ \text{s.t.} \quad g_i(\boldsymbol{x}) \leq 0, \quad i = 1, \ldots, l \\ \qquad \boldsymbol{x} \in \mathcal{X} = \{\boldsymbol{x} \in \mathbb{R}^n : x_j^{\min} \leq x_j \leq x_j^{\max},\ j = 1, \ldots, n\}, \end{cases}$$

where g_i, $i = 0, \ldots, l$, are continuously differentiable, g_0 is strictly convex and all other g_i are convex. In addition, g_i are separable, i.e. they may be written as a sum of functions of a single variable:

$$g_i(\boldsymbol{x}) = \sum_{j=1}^{n} g_{ij}(x_j), \quad i = 0, \ldots, l.$$

The separability of g_i makes it advantageous to use Lagrangian duality to solve the optimization problem. The Lagrangian function $\mathcal{L}$ of $(\mathbb{P})_s$ becomes

$$\begin{aligned}
\mathcal{L}(\boldsymbol{x}, \boldsymbol{\lambda}) &= g_0(\boldsymbol{x}) + \sum_{i=1}^{l} \lambda_i g_i(\boldsymbol{x}) \\
&= \sum_{j=1}^{n} g_{0j}(x_j) + \sum_{i=1}^{l} \lambda_i \left(\sum_{j=1}^{n} g_{ij}(x_j) \right) \\
&= \sum_{j=1}^{n} \underbrace{\left(g_{0j}(x_j) + \sum_{i=1}^{l} \lambda_i g_{ij}(x_j) \right)}_{\mathcal{L}_j(x_j, \boldsymbol{\lambda})},
\end{aligned}$$

where $\lambda_i \geq 0,\ i = 1, \ldots, l$. Note that $x_j \mapsto \mathcal{L}_j(x_j, \boldsymbol{\lambda})$ is strictly convex, cf. Lemma 3.1(ii). The dual objective function is

$$\varphi(\boldsymbol{\lambda}) = \min_{\boldsymbol{x}\in\mathcal{X}} \mathcal{L}(\boldsymbol{x}, \boldsymbol{\lambda}) = \min_{\boldsymbol{x}\in\mathcal{X}} \sum_{j=1}^{n} \mathcal{L}_j(x_j, \boldsymbol{\lambda}) = \sum_{j=1}^{n} \min_{\substack{x_j^{\min} \leq x_j \\ \leq x_j^{\max}}} \mathcal{L}_j(x_j, \boldsymbol{\lambda}).$$

Thus, in order to minimize the Lagrangian for all $\boldsymbol{x} \in \mathcal{X}$, we need only to perform n box constrained minimizations of functions of a single variable. This is the reason why Lagrangian duality is so attractive for convex, separable problems. In more detail, the minimization of $\mathcal{L}_j$ is performed as follows, when the lower and upper bounds are finite:

$$\begin{array}{llll} \text{if} & \dfrac{\partial \mathcal{L}_j(x_j^{\min}, \boldsymbol{\lambda})}{\partial x_j} \geq 0, & \text{then} & x_j^* = x_j^{\min} \\ \text{else if} & \dfrac{\partial \mathcal{L}_j(x_j^{\max}, \boldsymbol{\lambda})}{\partial x_j} \leq 0, & \text{then} & x_j^* = x_j^{\max} \\ \text{else} & & & x_j^* = x_j^*(\boldsymbol{\lambda}) \quad \text{from} \quad \dfrac{\partial \mathcal{L}_j(x_j, \boldsymbol{\lambda})}{\partial x_j} = 0, \end{array} \tag{3.14}$$

cf. Fig. 3.7.

Since $x_j \mapsto \mathcal{L}_j(x_j, \boldsymbol{\lambda})$ is strictly convex, there is a unique solution to this minimization problem. By performing (3.14) for each j, $1 \leq j \leq n$, we may obtain the dual objective function. As usual, the dual problem is solved by maximizing the dual objective function for $\boldsymbol{\lambda} \geq \boldsymbol{0}$.

Example 3.6 Consider the following convex and separable optimization problem, which we would like to solve using Lagrangian duality:

$$(\mathbb{P})_3 \quad \begin{cases} \min\limits_{x_1, x_2} (x_1 - 3)^2 + (x_2 + 1)^2 \\ \text{s.t.} \quad x_1 + x_2 - 1.5 \leq 0 \\ \qquad \boldsymbol{x} \in \mathcal{X} = \{\boldsymbol{x} : 0 \leq x_1 \leq 1, -2 \leq x_2 \leq 1\}. \end{cases}$$

The problem is illustrated in Fig. 3.8.

The Lagrangian function is

$$\begin{aligned} \mathcal{L}(\boldsymbol{x}, \lambda) &= (x_1 - 3)^2 + (x_2 + 1)^2 + \lambda(x_1 + x_2 - 1.5) \\ &= \underbrace{(x_1 - 3)^2 + \lambda x_1}_{\mathcal{L}_1(x_1, \lambda)} + \underbrace{(x_2 + 1)^2 + \lambda x_2 - 1.5\lambda}_{\mathcal{L}_2(x_2, \lambda)}. \end{aligned}$$

Differentiation gives

$$\frac{\partial \mathcal{L}_1}{\partial x_1} = 2x_1 + \lambda - 6, \qquad \frac{\partial \mathcal{L}_2}{\partial x_2} = 2x_2 + \lambda + 2. \tag{3.15}$$

Fig. 3.7 Minimization of $\mathcal{L}_j$

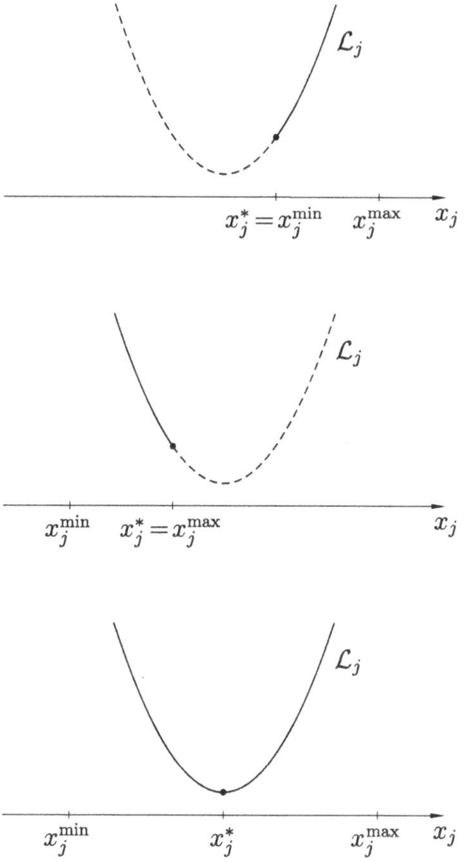

From (3.14), we find the $\boldsymbol{x}$, denoted $\boldsymbol{x}^*$, that minimizes $\mathcal{L}$ for any given $\lambda \geq 0$.

$$
\begin{aligned}
&\frac{\partial \mathcal{L}_1(0,\lambda)}{\partial x_1} = \lambda - 6 \geq 0 && \therefore x_1^* = 0, \text{ if } \lambda \geq 6 \\
&\frac{\partial \mathcal{L}_1(1,\lambda)}{\partial x_1} = \lambda - 4 \leq 0 && \therefore x_1^* = 1, \text{ if } 0 \leq \lambda \leq 4 \\
&\frac{\partial \mathcal{L}_1(x_1,\lambda)}{\partial x_1} = 2x_1 + \lambda - 6 = 0 && \therefore x_1^* = 3 - \frac{\lambda}{2}, \text{ if } 4 \leq \lambda \leq 6
\end{aligned}
\tag{3.16}
$$

$$
\begin{aligned}
&\frac{\partial \mathcal{L}_2(-2,\lambda)}{\partial x_2} = \lambda - 2 \geq 0 && \therefore x_2^* = -2, \text{ if } \lambda \geq 2 \\
&\frac{\partial \mathcal{L}_2(1,\lambda)}{\partial x_2} = \lambda + 4 \leq 0 && \text{never satisfied since } \lambda \geq 0 \\
&\frac{\partial \mathcal{L}_2(x_2,\lambda)}{\partial x_2} = 2x_2 + \lambda + 2 = 0 && \therefore x_2^* = -1 - \frac{\lambda}{2}, \text{ if } 0 \leq \lambda \leq 2.
\end{aligned}
\tag{3.17}
$$

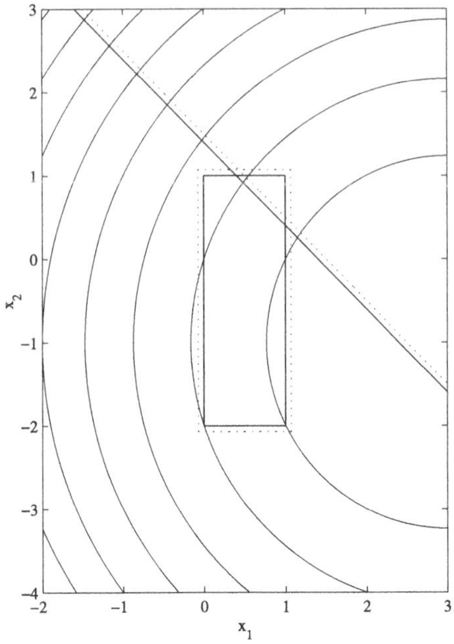

Fig. 3.8 A simple convex, separable problem

The dual objective function

$$\varphi(\lambda) = (x_1^* - 3)^2 + \lambda x_1^* + (x_2^* + 1)^2 + \lambda x_2^* - \frac{3}{2}\lambda \tag{3.18}$$

then becomes

$$\begin{cases} 4 + \lambda + \dfrac{\lambda^2}{4} - \lambda - \dfrac{\lambda^2}{2} - \dfrac{3}{2}\lambda = -\dfrac{\lambda^2}{4} - \dfrac{3}{2}\lambda + 4, & \text{if } 0 \le \lambda \le 2 \\ 4 + \lambda + 1 - 2\lambda - \dfrac{3}{2}\lambda = -\dfrac{5}{2}\lambda + 5, & \text{if } 2 \le \lambda \le 4 \\ \dfrac{\lambda^2}{4} + 3\lambda - \dfrac{\lambda^2}{2} + 1 - 2\lambda - \dfrac{3}{2}\lambda = -\dfrac{\lambda^2}{4} - \dfrac{\lambda}{2} + 1, & \text{if } 4 \le \lambda \le 6 \\ 9 + 1 - 2\lambda - \dfrac{3}{2}\lambda = -\dfrac{7}{2}\lambda + 10, & \text{if } \lambda \ge 6. \end{cases}$$

Note that φ is continuously differentiable ($\varphi(2) = 0$, $\varphi(4) = -5$, $\varphi(6) = -11$, $\varphi'(2) = -\frac{5}{2}$, $\varphi'(4) = -\frac{5}{2}$, $\varphi'(6) = -\frac{7}{2}$). The function is illustrated in Fig. 3.9. It is clear that φ is maximized for $\lambda^* = 0$. Insertion of this into (3.16) and (3.17) yields, according to Theorem 3.6, the optimal solution $x_1^* = 1$, $x_2^* = -1$ of $(\mathbb{P})_3$.

In general it is quite tedious to write φ as an explicit function of λ as done above. When solving small problems by hand it is usually more efficient to first assume that none of the box constraints is active. It may then turn out that the λ^* that maximizes φ results in a $\boldsymbol{x}^*$ that does not satisfy the box constraints. If this happens, one simply puts the primal variables $\boldsymbol{x}$ that do not satisfy the box constraints onto

Fig. 3.9 The dual objective function

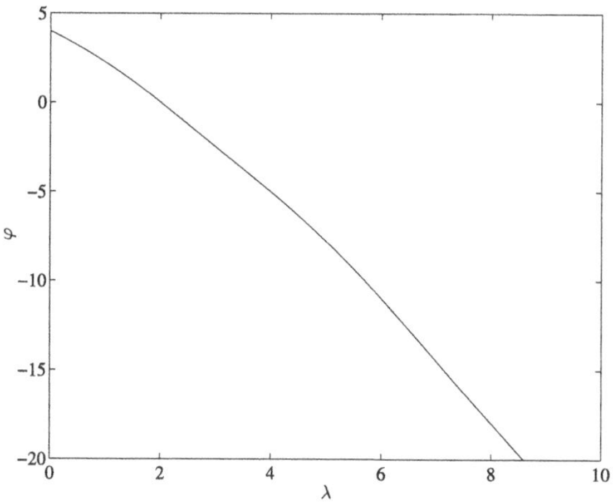

edges of the box region $\mathcal{X}$, and optimize again with these variables kept fixed. We illustrate the method for the problem at hand.

If the box constraints are not active, $\mathcal{L}_1$ and $\mathcal{L}_2$ are minimized when $\partial\mathcal{L}_1/\partial x_1 = 0$ and $\partial\mathcal{L}_2/\partial x_2 = 0$. Thus, (3.15) gives

$$x_1^* = 3 - \frac{\lambda}{2} \tag{3.19}$$

$$x_2^* = -1 - \frac{\lambda}{2}. \tag{3.20}$$

Insertion of this into (3.18) gives the dual objective function

$$\varphi(\lambda) = -\frac{\lambda^2}{2} + \frac{\lambda}{2},$$

which is maximized for $\lambda^* = \frac{1}{2}$. From (3.19) and (3.20) we then get $x_1^* = 11/4$ and $x_2^* = -5/4$. This cannot be the actual solution, however, since $x_1^* \nleq 1$. We therefore put $x_1^* = 1$ and try to find the optimum x_2^*. Again, $\partial\mathcal{L}_2/\partial x_2 = 0$ gives $x_2^* = -1 - \lambda/2$. The dual objective function is

$$\varphi(\lambda) = -\frac{\lambda^2}{4} - \frac{3}{2}\lambda + 4,$$

which is maximized for $\lambda^* = 0$. This results in the optimal solution $x_1^* = 1$ and $x_2^* = -1$ which is a feasible point. It is easily verified that the conditions in (3.14) are satisfied, so that $\mathcal{L}$ has really been minimized for $\boldsymbol{x}^*$. Since we have also maximized φ, the $\boldsymbol{x}^*$ obtained is the optimum solution of $(\mathbb{P})_3$.

Example 3.7 We will solve the two-bar truss problem on page 14 once again, this time using Lagrangian duality. From the primal problem $(\mathbb{P})_2$ on page 45, the Lagrangian function becomes

$$\mathcal{L}(A_1, A_2, \lambda) = \frac{2}{\sqrt{3}} A_1 + A_2 + \lambda \left(\frac{8}{\sqrt{3} A_1} + \frac{3}{A_2} - \frac{\sigma_0}{F} \right). \tag{3.21}$$

We differentiate $\mathcal{L}$ and assume that the box constraints are not active:

$$\frac{\partial \mathcal{L}}{\partial A_1} = \frac{2}{\sqrt{3}} - \frac{8}{\sqrt{3} A_1^2} \lambda = 0,$$

$$\frac{\partial \mathcal{L}}{\partial A_2} = 1 - \frac{3}{A_2^2} \lambda = 0,$$

which gives

$$A_1 = 2\sqrt{\lambda}, \quad A_2 = \sqrt{3\lambda}. \tag{3.22}$$

The dual objective function is obtained by insertion of these expressions into (3.21) as

$$\begin{aligned} \varphi(\lambda) &= \frac{4}{\sqrt{3}} \sqrt{\lambda} + \sqrt{3\lambda} + \lambda \left(\frac{4}{\sqrt{3}\sqrt{\lambda}} + \frac{3}{\sqrt{3\lambda}} - \frac{\sigma_0}{F} \right) \\ &= \frac{14}{\sqrt{3}} \sqrt{\lambda} - \frac{\sigma_0}{F} \lambda. \end{aligned}$$

Maximization of φ for $\lambda \geq 0$ yields

$$\sqrt{\lambda} = \frac{7F}{\sqrt{3}\sigma_0}.$$

Insertion of this into (3.22) gives the optimum solution as it is easily checked that the box constraints indeed are not active or violated. A final comment: in this problem the objective function is linear and thus convex, but not strictly convex. The reason that we nevertheless obtain a unique solution of (3.14) is that $x_j \mapsto \mathcal{L}_j(x_j, \boldsymbol{\lambda})$ is still strictly convex since g_1 in $(\mathbb{P})_2$ is strictly convex and the corresponding Lagrangian multiplier $\lambda > 0$.

3.5 Exercises

Exercise 3.1 Show that problem $(\mathbb{SO})^3_{\mathrm{nf}}$ on page 17 and problem $(\mathbb{SO})^4_{\mathrm{nf}}$ on page 20 are convex.

Exercise 3.2 Show that the problem of Exercise 2.5 is nonconvex.

Exercise 3.3 One wants to minimize the weight of the two-bar truss in Fig. 3.10. The lengths of the bars are $5l$ and $3l$, respectively. Young's modulus is E and the density is ρ for both bars, and the force $P > 0$. The design variables are the cross-sectional areas of the bars: A_1 and A_2. The truss has to be sufficiently stiff; more precisely, the so-called compliance has to be lower than a specified number:

$$-Pu_x - Pu_y \leq c_0,$$

where (u_x, u_y) are the displacements of the free node, and $c_0 > 0$ is a given number.

a) Formulate the problem as a mathematical programming problem.

b) Change variables to nondimensional ones as $x_i = P/(EA_i)$, $i = 1, 2$, and solve the optimization problem by using the KKT conditions.

c) Same as b), but solve the optimization problem by using Lagrangian duality instead.

Exercise 3.4 The three-bar truss in Fig. 3.11 is subjected to the force $P>0$. One wants to maximize the stiffness of the truss by minimizing its compliance Pu_y, where u_y is the displacement in the y-direction of the node where P is applied. The volume of the truss may not exceed the value V_0. The design variables are the cross-sectional areas of the bars: A_1, A_2 and A_3.

a) Formulate the problem as a mathematical programming problem.
b) Solve the optimization problem by using the KKT conditions.
c) Solve the optimization problem by using Lagrangian duality.

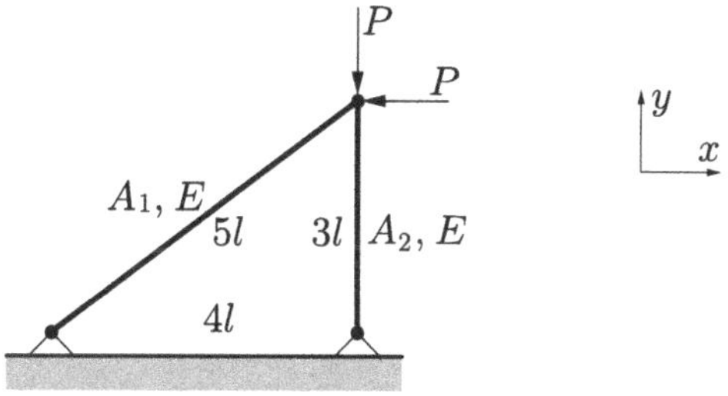

Fig. 3.10 The two-bar truss of Exercise 3.3

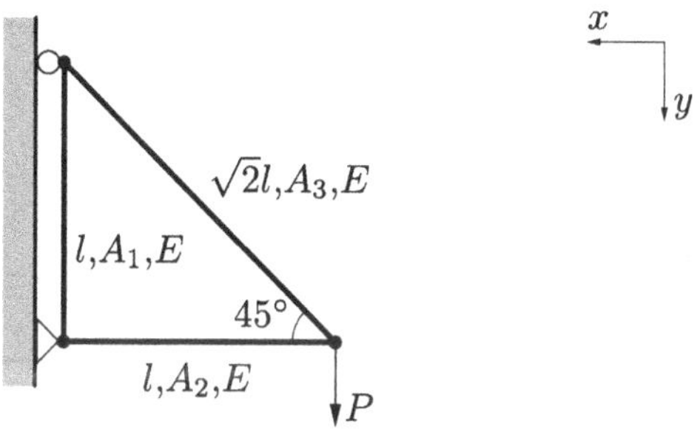

Fig. 3.11 The three-bar truss of Exercise 3.4

Exercise 3.5 The two-bar truss in Fig. 3.12 is subjected to the force $P > 0$. The compliance $-Pu_y$ should be minimized, where u_y is the displacement in the y-direction of the free node. The volume of the truss may not exceed the value V_0, and the magnitude of the stress in each bar (both in tension and compression) is not allowed to exceed the value $(5\alpha Pl)/(6V_0)$, where $\alpha > 0$ is a given dimensionless constant. The design variables are the cross-sectional areas of the bars: A_1 and A_2.

a) Formulate the problem as a mathematical programming problem.
b) Solve the optimization problem by using Lagrangian duality for all $\alpha > 0$.

Exercise 3.6 The stiffness of the three-bar truss in Fig. 3.13 should be maximized. More precisely, one wants to minimize the 1-norm of the displacement vector, i.e.

$$|u_{1_x}| + |u_{1_y}| + |u_{2_x}|.$$

The truss is subjected to two forces $P > 0$. The volume of the truss may not exceed the value V_0. The design variables are the cross-sectional areas of the bars: A_1, A_2 and A_3.

a) Formulate the problem as a mathematical programming problem.
b) Solve the optimization problem by using Lagrangian duality.

Exercise 3.7 The weight of the three-bar truss in Fig. 3.14 should be minimized given that the truss should be sufficiently stiff; the maximum nodal displacement

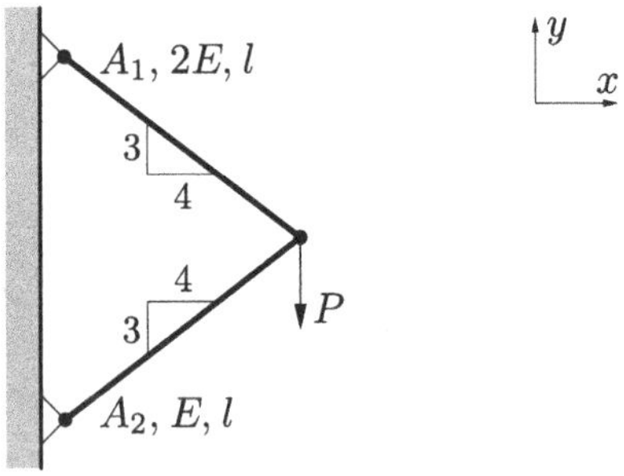

Fig. 3.12 The two-bar truss of Exercise 3.5

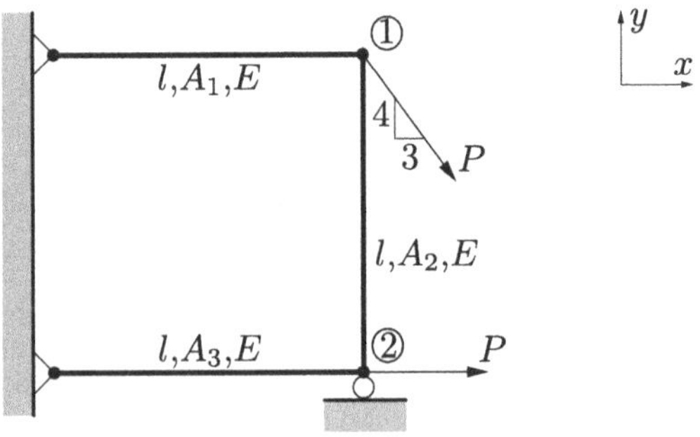

Fig. 3.13 The three-bar truss of Exercise 3.6

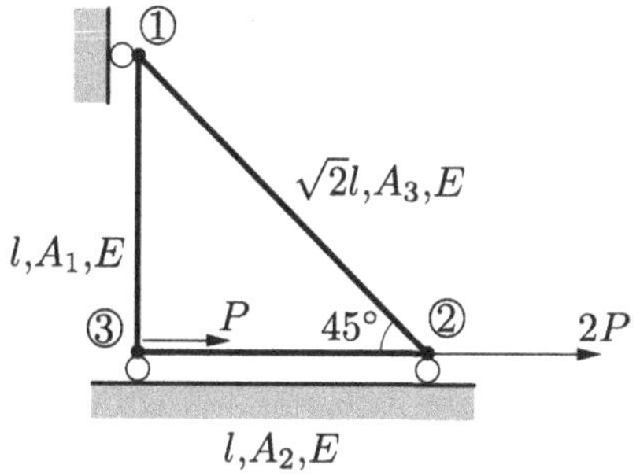

Fig. 3.14 The three-bar truss of Exercise 3.7

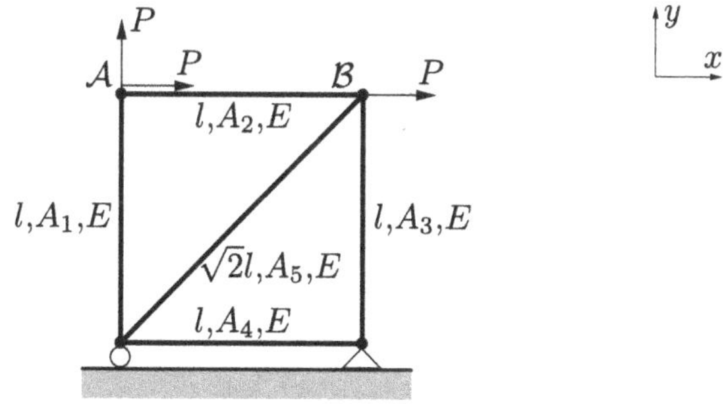

Fig. 3.15 The five-bar truss of Exercise 3.8

has to be lower than a prescribed value:

$$\max(|\boldsymbol{u}_1|, |\boldsymbol{u}_2|, |\boldsymbol{u}_3|) \leq u_0,$$

where $\boldsymbol{u}_i$ is the displacement vector of node i and $u_0 > 0$ is a given scalar. The truss is subjected to two applied forces. It holds that $P > 0$. The design variables are the cross-sectional areas of the bars: A_1, A_2 and A_3.

a) Formulate the problem as a mathematical programming problem.
b) Solve the optimization problem by using Lagrangian duality.

Exercise 3.8 The volume of the five-bar truss in the Fig. 3.15 should be minimized given that the truss has to be sufficiently stiff. Specifically, the compliance has to be lower than the value c_0. All bars have Young's modulus E. The design variables are the cross-sectional areas of the bars: $A_1, A_2, \ldots, A_5$. The truss is subjected to three forces $P > 0$, so that the compliance may be written as

$$Pu_x^{\mathcal{A}} + Pu_y^{\mathcal{A}} + Pu_x^{\mathcal{B}},$$

where $(u_x^{\mathcal{N}}, u_y^{\mathcal{N}})$ are the displacements of the node $\mathcal{N}$.

a) Formulate the problem as a mathematical programming problem.
b) Solve the optimization problem by using Lagrangian duality.

Chapter 4
Sequential Explicit, Convex Approximations

In the previous two chapters we were able to formulate a number of structural optimization problems where both the objective function and all of the constraints were written as explicit functions of the design variables only. For larger problems, however, it is in general practically impossible to obtain such explicit functions. Our remedy to be able to solve large-scale problems is to generate a sequence of explicit subproblems that are approximations of the original problem and solve these subproblems instead.

As already mentioned, most problems in structural optimization are nonconvex. Because of the intrinsic difficulties with solving nonconvex problems, we will choose approximations that are convex. In this chapter, a number of explicit, convex approximations will be described. The main focus will be on approximations that take into account specific characteristics of certain structural optimization problems.

4.1 General Solution Procedure for Nested Problems

Let us study a structural optimization problem of a system with a finite number of degrees-of-freedom, such as a truss. If linear elasticity is assumed, the problem under study is written as follows, using a simultaneous formulation:

$$(\mathbb{SO})_{\text{sf}} \quad \begin{cases} \min\limits_{\boldsymbol{x},\boldsymbol{u}} g_0(\boldsymbol{x},\boldsymbol{u}) \\ \text{s.t.} \quad \boldsymbol{K}(\boldsymbol{x})\boldsymbol{u} = \boldsymbol{F}(\boldsymbol{x}) \\ \qquad g_i(\boldsymbol{x},\boldsymbol{u}) \leq 0, \ i = 1, \ldots, l \\ \qquad \boldsymbol{x} \in \mathcal{X} = \{\boldsymbol{x} \in \mathbb{R}^n x_j^{\min} \leq x_j \leq x_j^{\max}, \ j = 1, \ldots, n\}, \end{cases}$$

where $\boldsymbol{K}(\boldsymbol{x})$ is the global stiffness matrix of the structure, $\boldsymbol{u}$ is the global displacement vector, and $\boldsymbol{F}(\boldsymbol{x})$ is the global external force vector. It is certainly possible to solve $(\mathbb{SO})_{\text{sf}}$ directly, but there is major disadvantage with the simultaneous formulation for large-scale problems—the number of constraints due to the equilibrium equations is huge. In case the stiffness matrix is nonsingular, we may use the equilibrium equations to write the displacements as functions of the design variables: $\boldsymbol{u}(\boldsymbol{x}) = \boldsymbol{K}^{-1}(\boldsymbol{x})\boldsymbol{F}(\boldsymbol{x})$. Now, for small problems we can easily obtain $\boldsymbol{u}(\boldsymbol{x})$ explicitly, i.e. as a symbolic formula, see for instance (2.32)–(2.33). For larger problems it would be extremely time-consuming to produce such formulas. In this case, the equilibrium equations will be used to *implicitly* define $\boldsymbol{u}(\boldsymbol{x})$. That is, we do not write $\boldsymbol{u}(\boldsymbol{x})$ as an explicit formula, but only take advantage of the fact that $\boldsymbol{K}(\boldsymbol{x})$, $\boldsymbol{u}(\boldsymbol{x})$ and $\boldsymbol{F}(\boldsymbol{x})$ are related through the formula $\boldsymbol{K}(\boldsymbol{x})\boldsymbol{u}(\boldsymbol{x}) = \boldsymbol{F}(\boldsymbol{x})$. Although it is practically impossible to write $\boldsymbol{u}(\boldsymbol{x})$ as an explicit function for large problems, it is always possible to solve the equilibrium equations numerically for $\boldsymbol{u}(\bar{x})$ for any given design $\bar{x}$.

By using the equilibrium equations to write the displacements as functions of the design variables, we obtain the nested formulation of the structural optimization problem as

$$(\mathbb{SO})_{\text{nf}} \quad \begin{cases} \min\limits_{x} \hat{g}_0(\boldsymbol{x}) \\ \text{s.t.} \quad \hat{g}_i(\boldsymbol{x}) \leq 0, \quad i = 1, \ldots, l \\ \qquad \boldsymbol{x} \in \mathcal{X}, \end{cases}$$

where $\hat{g}_i(\boldsymbol{x}) = g_i(\boldsymbol{x}, \boldsymbol{u}(\boldsymbol{x}))$, $i = 0, \ldots, l$. The problem $(\mathbb{SO})_{\text{nf}}$ will be solved by generating and solving a sequence of explicit subproblems that are approximations of $(\mathbb{SO})_{\text{nf}}$. The optimization algorithms used to solve the subproblems will obviously need information about $\hat{g}_i$, $i = 0, \ldots, l$, and, possibly, their derivatives. We say that an algorithm is of *order* j if the highest order of derivatives used is j. In structural optimization, first order methods are most common. Zero order methods, which consequently do not use any derivatives, but only need $\hat{g}_i$, $i = 0, \ldots, l$, to be calculated, have begun to attract some interest in recent years, at least for small-scale problems. Second or higher order methods are rarely used as the calculation of higher order derivatives is expensive.

The procedure for solving the nested formulation of the structural optimization problem using a first order algorithm may be described in the following steps:

1. Start with an initial design $\boldsymbol{x}^0$. Set the iteration counter $k = 0$.
2. Calculate the displacement vector $\boldsymbol{u}(\boldsymbol{x}^k)$ for the current design by performing a finite element analysis: $\boldsymbol{K}(\boldsymbol{x}^k)\boldsymbol{u}(\boldsymbol{x}^k) = \boldsymbol{F}(\boldsymbol{x}^k)$.
3. For the current design $\boldsymbol{x}^k$, calculate the objective function $\hat{g}_0(\boldsymbol{x}^k)$, the constraint functions $\hat{g}_i(\boldsymbol{x}^k)$, $i = 1, \ldots, l$, and their gradients $\nabla \hat{g}_i(\boldsymbol{x}^k)$, $i = 0, \ldots, l$.
4. Formulate an explicit, convex approximation $(\mathbb{SO})^k_{\text{nf}}$ at $\boldsymbol{x}^k$ of $(\mathbb{SO})_{\text{nf}}$.
5. Solve $(\mathbb{SO})^k_{\text{nf}}$ by a nonlinear optimization algorithm to give a new design $\boldsymbol{x}^{k+1}$.
6. Put $k = k + 1$ and return to step 2 unless a stopping criterion is satisfied.

In the next sections, we will describe four different methods to obtain an explicit approximation of $(\mathbb{SO})_{\text{nf}}$: SLP, SQP, CONLIN and MMA. For notational convenience we will skip the circumflex (hat) in $\hat{g}_i$, $i = 0, \ldots, l$.

We end this section by remarking that the numerical efficiency of an optimization algorithm is in general improved if one scales the variables and constraints so that they are of a similar order of magnitude. In Chap. 7, for instance, we will use design variables that all lie between 0 and 1 in order to vary the shape of a structure.

4.2 Sequential Linear Programming (SLP)

In a Sequential Linear Programming (SLP) approximation of $(\mathbb{SO})_{\text{nf}}$ at $\boldsymbol{x}^k$, the objective function and all constraint equations are linearized at the design $\boldsymbol{x}^k$; this leads to the following subproblem at iteration k:

$$
\text{(SLP)} \quad \begin{cases} \min\limits_{\mathbf{x}} g_0(\boldsymbol{x}^k) + \nabla g_0(\boldsymbol{x}^k)^T(\boldsymbol{x} - \boldsymbol{x}^k) \\ \text{s.t.} \quad g_i(\boldsymbol{x}^k) + \nabla g_i(\boldsymbol{x}^k)^T(\boldsymbol{x} - \boldsymbol{x}^k) \leq 0, \quad i = 1, \ldots, l \\ \qquad \mathbf{x} \in \mathcal{X} \\ \qquad -l_j^k \leq x_j - x_j^k \leq u_j^k, \quad j = 1, \ldots, n. \end{cases}
$$

Here, l_j^k and u_j^k, $j = 1, \ldots, n$, are so-called move limits; these are used since in general the linearization used is accurate only close to the current design. The move limits are updated according to some user-defined rule during the iterations. It turns out that the choice of move limits affects the efficiency of (SLP) considerably.

Once $g_i(\boldsymbol{x}^k)$ and $\nabla g_i(\boldsymbol{x}^k)$, $i = 0, \ldots, l$, have been calculated, all expressions in (SLP) are known, explicit functions of $\boldsymbol{x}$. Thus, (SLP) is indeed an *explicit* approximation of $(\mathbb{SO})_{\text{nf}}$.

Since the objective function and all constraints in (SLP) are affine functions of $\boldsymbol{x}$, i.e. they may be written on the form $\boldsymbol{a}^T\boldsymbol{x} + b$, where $\boldsymbol{a}$ and b are constants, they are convex. Consequently, (SLP) is a convex problem. Since all g_i, $i = 0, \ldots, l$, are affine, (SLP) is a Linear Problem (LP), which may be solved by, e.g., the Simplex algorithm.

4.3 Sequential Quadratic Programming (SQP)

By adding a second order term in the Taylor expansion of the objective function in (SLP), one obtains a Sequential Quadratic Programming (SQP) approximation:

$$
\text{(SQP)} \quad \begin{cases} \min\limits_{\boldsymbol{x}} g_0(\boldsymbol{x}^k) + \nabla g_0(\boldsymbol{x}^k)^T(\boldsymbol{x} - \boldsymbol{x}^k) + \frac{1}{2}(\boldsymbol{x} - \boldsymbol{x}^k)^T \boldsymbol{H}(\boldsymbol{x}^k)(\boldsymbol{x} - \boldsymbol{x}^k) \\ \text{s.t.} \quad g_i(\boldsymbol{x}^k) + \nabla g_i(\boldsymbol{x}^k)^T(\boldsymbol{x} - \boldsymbol{x}^k) \leq 0, \quad i = 1, \ldots, l \\ \qquad \boldsymbol{x} \in \mathcal{X}, \end{cases}
$$

where $\boldsymbol{H}(\boldsymbol{x}^k)$ denotes a positive definite, first order approximation of the Hessian of g_0 at $\boldsymbol{x}^k$. This will lead to a convex objective function, so that (SQP) becomes a convex problem. In a general SQP implementation, $\boldsymbol{H}$ may very well be chosen as the actual Hessian, but we do not use it here, since we will use only first order methods. We have opted not to include move limits, as their importance is far less than for (SLP) since (SQP) is a better approximation of the original problem at the current design.

4.4 Convex Linearization (CONLIN)

Both SLP and SQP are designed to solve general nonlinear optimization problems of the form $(\mathbb{SO})_{\text{nf}}$, i.e. they do not take into account any specific characteristics that structural optimization problems might have. In order to try to find such special characteristics, let us return to Sect. 2.3 where the cross-sectional areas A_i, $i = 1, 2$, of the bars that minimized the weight of a statically determinate two-bar truss under stress and displacement constraints were sought. It was found that the stresses can be written as

$$\sigma_i = \frac{b_i}{A_i}, \quad i = 1, 2,$$

for some constants b_i, whereas the displacements become

$$u_i = \sum_{j=1}^{2} \frac{b_{ij}}{A_j}, \quad i = 1, 2,$$

where $u_1 = u_x$, $u_2 = u_y$ and b_{ij} are constants. Thus, stresses and displacements are functions of $1/A_i$. We will see in the next chapter that this holds for any statically determinate truss. For a statically indeterminate truss, however, the expressions for the stresses and displacements are more complicated; cf. the expressions (2.32)–(2.36) for the three-bar truss in Sect. 2.5.

Our conclusion from this investigation is that if we were to linearize the stress and displacement constraints in the so-called *reciprocal variables* $1/A_i$, then these linearizations would be exact for statically determinate trusses. For other trusses, the linearizations would not be exact, but it seems reasonable to expect that it is better to linearize in $1/A_i$ rather than in the direct variables A_i as done in SLP and SQP.

It would be a bad idea to linearize every possible objective function or constraint function in the variables $1/A_i$, however. For instance, the weight in this example is already a linear function of A_i. An idea springs to mind: linearize some functions in the variables A_i and others in the variables $1/A_i$. This is exactly what is done in the approximation method CONLIN (Convex Linearization), developed by Fleury [15].

In CONLIN, one assumes that all design variables are strictly positive, i.e. the set $\mathcal{X}$ in $(\mathbb{SO})_{\text{nf}}$ is changed to

$$\mathcal{X} = \{\boldsymbol{x} \in \mathbb{R}^n \, : \, 0 < x_j^{\min} \leq x_j \leq x_j^{\max}, \; j = 1, \ldots, n\}.$$

The objective function $g_0(\boldsymbol{x})$ and all constraint functions $g_i(\boldsymbol{x})$, $i = 1, \ldots, l$, are linearized at the design $\boldsymbol{x}^k$ in the *intervening variables* $y_j = y_j(x_j)$, $j = 1, \ldots, n$, where y_j will be either x_j or $1/x_j$:

$$g_i(\boldsymbol{x}) \approx g_i(\boldsymbol{x}^k) + \sum_{j=1}^{n} \frac{\partial g_i(\boldsymbol{x}^k)}{\partial y_j} \left(y_j(x_j) - y_j(x_j^k) \right). \tag{4.1}$$

The partial derivative of g_i with respect to the intervening variable y_j is obtained by employing the chain rule as

$$\frac{\partial g_i(\boldsymbol{x}^k)}{\partial y_j} = \frac{\partial g_i(\boldsymbol{x}^k)}{\partial x_j} \frac{dx_j(x_j^k)}{dy_j} = \frac{\partial g_i(\boldsymbol{x}^k)}{\partial x_j} \frac{1}{\frac{dy_j(x_j^k)}{dx_j}}.$$

Next, we determine the contribution to the sum in (4.1) for the case $y_j = x_j$ and $y_j = 1/x_j$, respectively. Choosing $y_j = x_j$ gives us

$$g_{ij}^{L,k}(\boldsymbol{x}) = \frac{\partial g_i(\boldsymbol{x}^k)}{\partial x_j}\left(x_j - x_j^k\right), \tag{4.2}$$

whereas for $y_j = 1/x_j$,

$$g_{ij}^{R,k}(\boldsymbol{x}) = \frac{\partial g_i(\boldsymbol{x}^k)}{\partial x_j}\frac{1}{\left(-\frac{1}{(x_j^k)^2}\right)}\left(\frac{1}{x_j} - \frac{1}{x_j^k}\right) = \frac{\partial g_i(\boldsymbol{x}^k)}{\partial x_j}\frac{x_j^k(x_j - x_j^k)}{x_j}. \tag{4.3}$$

We define the following approximation of g_i at $\boldsymbol{x}^k$:

$$g_i^{RL,k}(\boldsymbol{x}) = g_i(\boldsymbol{x}^k) + \sum_{j \in \Omega_L} g_{ij}^{L,k}(\boldsymbol{x}) + \sum_{j \in \Omega_R} g_{ij}^{R,k}(\boldsymbol{x}), \tag{4.4}$$

where $\Omega_L = \{j : y_j = x_j\}$ and $\Omega_R = \{j : y_j = 1/x_j\}$. We need to supply a rule to decide what variables should be linearized in the direct variables x_j and what variables should be linearized in the reciprocal variables $1/x_j$. In CONLIN, the approximation of g_i at $\boldsymbol{x}^k$ is defined to be

$$g_i^{C,k}(\boldsymbol{x}) = g_i(\boldsymbol{x}^k) + \sum_{j \in \Omega_+} g_{ij}^{L,k}(\boldsymbol{x}) + \sum_{j \in \Omega_-} g_{ij}^{R,k}(\boldsymbol{x}), \tag{4.5}$$

where

$$\Omega^+ = \{j : \partial g_i(\boldsymbol{x}^k)/\partial x_j > 0\} \quad \text{and} \quad \Omega^- = \{j : \partial g_i(\boldsymbol{x}^k)/\partial x_j \leq 0\}.$$

That is, one linearizes in the direct variables if the corresponding component of the gradient is positive, and in the reciprocal variables otherwise. The CONLIN approximation turns out to be the most conservative approximation that can be obtained for an approximation on the form (4.4). This means that $g_i^{C,k}(\boldsymbol{x}) \geq g_i^{RL,k}(\boldsymbol{x})$ for every possible choice of the sets Ω_L and Ω_R, i.e. for every possible choice deciding which variables should be linearized in the direct and reciprocal variables, respectively.

The reason why $g_i^{C,k}(\boldsymbol{x})$ is more conservative than $g_i^{RL,k}(\boldsymbol{x})$ simply because $g_i^{C,k}(\boldsymbol{x}) \geq g_i^{RL,k}(\boldsymbol{x})$ is that it is more conservative to choose the larger objective function since we are solving a minimization problem, and the feasible set $\{\boldsymbol{x} \in \mathcal{X} : g_i^{C,k}(\boldsymbol{x}) \leq 0,\ i = 1, \ldots, l\}$ will be smaller than $\{\boldsymbol{x} \in \mathcal{X} : g_i^{RL,k}(\boldsymbol{x}) \leq 0,\ i = 1, \ldots, l\}$. For an illustration, study the following two optimization problems:

$$(\mathbb{P})_1 \quad \begin{cases} \min\limits_{x} g_0(x) \\ \text{s.t.} \quad g_1(x) \leq 0 \end{cases} \qquad (\mathbb{P})_2 \quad \begin{cases} \min\limits_{x} \bar{g}_0(x) \\ \text{s.t.} \quad \bar{g}_1(x) \leq 0, \end{cases}$$

where $g_0(x) \leq \bar{g}_0(x)$ and $g_1(x) \leq \bar{g}_1(x)$, see Fig. 4.1. Since $\bar{g}_0$ is more conservative than g_0 and $\bar{g}_1$ is more conservative than g_1, the solution of $(\mathbb{P})_2$ is more conservative, i.e. larger, than that of $(\mathbb{P})_1$.

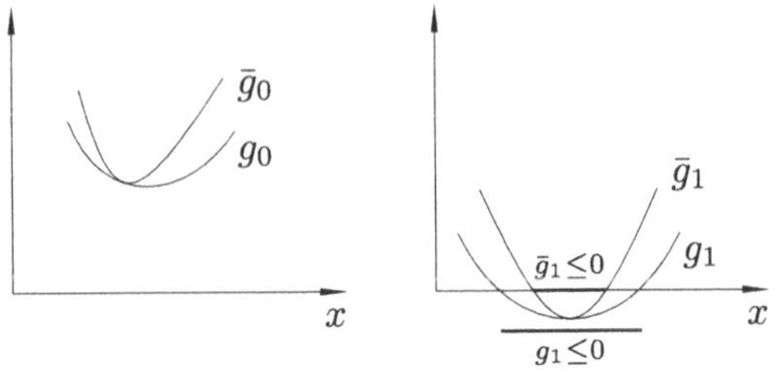

Fig. 4.1 $\bar{g}_i$ is more conservative than g_i, $i = 0, 1$

We will now prove that $g_i^{C,k}(\boldsymbol{x}) \geq g_i^{RL,k}(\boldsymbol{x})$. We have that

$$g_i^{C,k}(\boldsymbol{x}) - g_i^{RL,k}(\boldsymbol{x}) = \sum_{j \in \Omega_+ \cap \Omega_R} \left(g_{ij}^{L,k}(\boldsymbol{x}) - g_{ij}^{R,k}(\boldsymbol{x}) \right) + \sum_{j \in \Omega_- \cap \Omega_L} \left(g_{ij}^{R,k}(\boldsymbol{x}) - g_{ij}^{L,k}(\boldsymbol{x}) \right), \tag{4.6}$$

where $g_{ij}^{L,k}(\boldsymbol{x}) - g_{ij}^{R,k}(\boldsymbol{x})$ is obtained from (4.2) and (4.3) as

$$g_{ij}^{L,k}(\boldsymbol{x}) - g_{ij}^{R,k}(\boldsymbol{x}) = \frac{\partial g_i(\boldsymbol{x}^k)}{\partial x_j} \left(x_j - x_j^k - \frac{x_j^k (x_j - x_j^k)}{x_j} \right) = \frac{\partial g_i(\boldsymbol{x}^k)}{\partial x_j} \frac{x_j - x_j^k}{x_j} (x_j - x_j^k).$$

Since $x_j > 0$ according to the definition of the set $\mathcal{X}$, all terms in the sums in (4.6) are nonnegative, and thus $g_i^{C,k}(\boldsymbol{x}) \geq g_i^{RL,k}(\boldsymbol{x})$.

We have designed $g_i^{C,k}$ such that it is always more, or equally, conservative than the linear approximation

$$g_i^{L,k}(\boldsymbol{x}) = g_i(\boldsymbol{x}^k) + \sum_{j=1}^{n} \frac{\partial g_i(\boldsymbol{x}^k)}{\partial x_j} \left(x_j - x_j^k \right).$$

That is, CONLIN is more conservative than SLP. Note however, that $g_i^{C,k}$ may very well be less conservative than the original function g_i.

Some important properties of the CONLIN approximation follow:

- $g_i^{C,k}$ is a first order approximation of g_i, i.e. the function values and the first order partial derivatives are exact at $\boldsymbol{x} = \boldsymbol{x}^k$: $g_i^{C,k}(\boldsymbol{x}^k) = g_i(\boldsymbol{x}^k)$ and $\partial g_i^{C,k}(\boldsymbol{x}^k)/\partial x_j = \partial g_i(\boldsymbol{x}^k)/\partial x_j$.

- $g_i^{C,k}$ is an explicit, convex approximation. The convexity follows by noting that for each j, the contribution to $g_i^{C,k}$ is either $x_j - x_j^k$ times the gradient at $\boldsymbol{x}^k$ or

$$\frac{\partial g_i(\boldsymbol{x}^k)}{\partial x_j}\frac{x_j^k}{x_j}(x_j - x_j^k) = \frac{\partial g_i(\boldsymbol{x}^k)}{\partial x_j}\left(x_j^k - \frac{(x_j^k)^2}{x_j}\right).$$

  This last expression is valid when the partial derivative of g_i with respect to x_j is negative, which implies that the expression may be written as $A + B/x_j$ where A and $B > 0$ are constants. The terms $C(x_j - x_j^k)$, where C is a constant, and $A + B/x_j$ are both convex. Since $g_i^{C,k}$ is obtained by summing such terms and adding a constant, it follows from Lemma 3.1(ii) that the CONLIN approximation g_i^C is indeed a convex function as the name indicates!
- $g_i^{C,k}$ is a separable approximation since there obviously exist functions g_{ij} such that $g_i^{C,k}(\boldsymbol{x}) = \sum_{j=1}^n g_{ij}(x_j)$.

The fact that CONLIN is a convex, separable approximation makes Lagrangian duality a suitable solution method for the approximation of $(\mathbb{SO})_{\mathrm{nf}}$ at iteration k:

$$(\text{CONLIN}) \quad \begin{cases} \min\limits_{\boldsymbol{x}} g_0^{C,k}(\boldsymbol{x}) \\ \text{s.t.} \quad g_i^{C,k}(\boldsymbol{x}) \le 0, \quad i = 1, \dots, l \\ \qquad\; \boldsymbol{x} \in \mathcal{X}. \end{cases}$$

In practice, a term $\alpha \sum_{j=1}^n (x_j - x_j^k)^2$, where α is a small positive number is added to the objective function to ensure that it is strictly convex.

Example 4.1 Consider the function $g(x) = x + x^2 - \frac{1}{40}x^4$. We would like to calculate the CONLIN approximations of this function at $\bar{x} = 1$ and $\bar{\bar{x}} = 6$.

We start by differentiating g: $g_x(x) = \frac{\partial g(x)}{\partial x} = 1 + 2x - \frac{1}{10}x^3$. Thus, $g(\bar{x}) = 1.975$ and $g_x(\bar{x}) = 2.9 > 0$, so that the CONLIN approximation becomes the linear approximation: $g^C(x) = g^L(x) = 1.975 + 2.9(x - 1)$. In Fig. 4.2, g, g^C as well as the reciprocal approximation $g^R(x) = 1.975 + \frac{2.9}{x}(x - 1)$ are plotted. As always, g^C is larger than, or equal to, g^L and g^R. It is seen that in a neighborhood of $\bar{x}$, g is larger than g^C which illustrates the fact that the CONLIN approximation need not be more conservative than the original function. For the point $\bar{\bar{x}} = 6$, we have $g(\bar{\bar{x}}) = 9.6$, $g_x(\bar{\bar{x}}) = -8.6 < 0$ so that the CONLIN approximation now becomes the reciprocal approximation: $g^C(x) = g^R(x) = 9.6 + \frac{51.6}{x}(x - 6)$. For comparison, we also calculate the linear approximation as $g^L(x) = 9.6 - 8.6(x - 6)$, see Fig. 4.2.

Example 4.2 The volume of the four-bar truss in Fig. 4.3 should be minimized under the constraint that the elongation of bar 1, $\delta \le \delta_0$. The external force P is positive. The bars all have length l and Young's modulus E. It holds that $A_3 = A_2$ and $A_4 = A_1$. The design variables are the cross-sectional areas A_1 and A_2. We define $A_0 = Pl/(10\delta_0 E)$. There are bounds on A_1 and A_2 according to

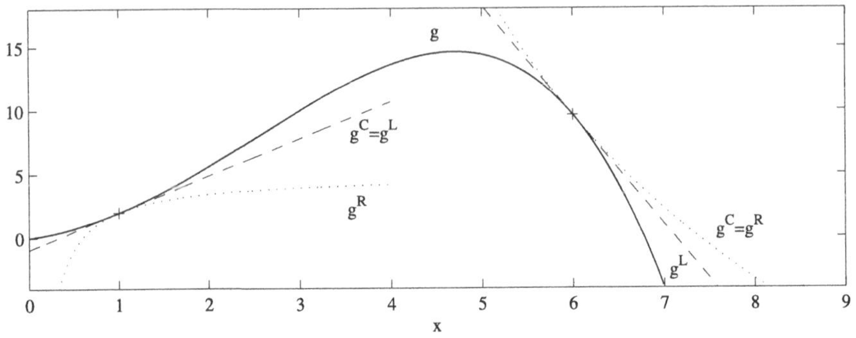

Fig. 4.2 CONLIN approximations of a function g

Fig. 4.3 A four-bar truss

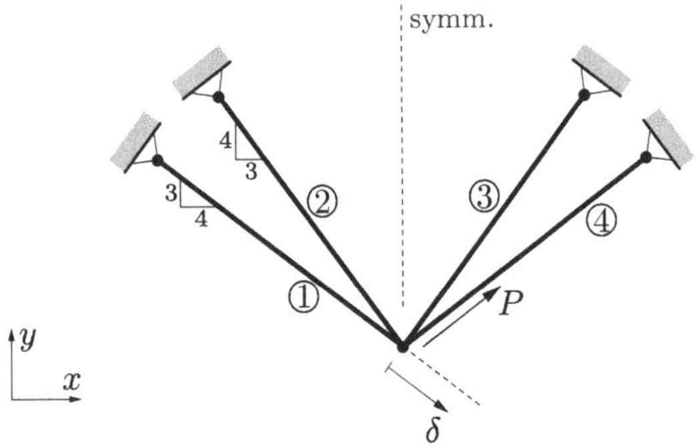

$0.2A_0 \leq A_i \leq 2.5A_0,\ i=1,2$. New variables are introduced as $x_i = A_i/A_0,\ i = 1,2$. The optimization problem may be written as

$$(\mathbb{P})_3 \quad \begin{cases} \min\limits_{x_1,x_2} \ g_0(x_1,x_2) = x_1 + x_2 \\ \text{s.t.} \quad \begin{cases} g_1(x_1,x_2) = \dfrac{8}{16x_1+9x_2} - \dfrac{4.5}{9x_1+16x_2} - 0.1 \leq 0 \\ 0.2 \leq x_1 \leq 2.5, \qquad 0.2 \leq x_2 \leq 2.5. \end{cases} \end{cases}$$

This problem, like most problems in structural optimization, is nonconvex, see Fig. 4.4. It may be solved by generating a sequence of convex CONLIN subproblems. The initial design is chosen as $\boldsymbol{x}^0 = (2,1)$. The CONLIN approximation of the objective function will be identical to the function itself. The approximation of the constraint function g_1 at $\boldsymbol{x}^0$ is illustrated in the figure. It is seen that the CONLIN approximated subproblem is convex. If this subproblem is solved, for instance by use of Lagrangian duality, the point $\boldsymbol{x}^1 = (1.2, 0.2)$ is obtained. This point is obviously not the solution of the original problem $(\mathbb{P})_3$. A new iteration is therefore performed. The CONLIN approximation of the constraint function at $\boldsymbol{x}^1$ is calculated, and the subproblem obtained is solved to yield a new design $\boldsymbol{x}^2 = (0.85, 0.2)$. It turns out that this point is located slightly outside the set $\{(x_1,x_2) : g_1(x_1,x_2) \leq 0\}$, which, again, confirms that the CONLIN approximation of a function may very well be less

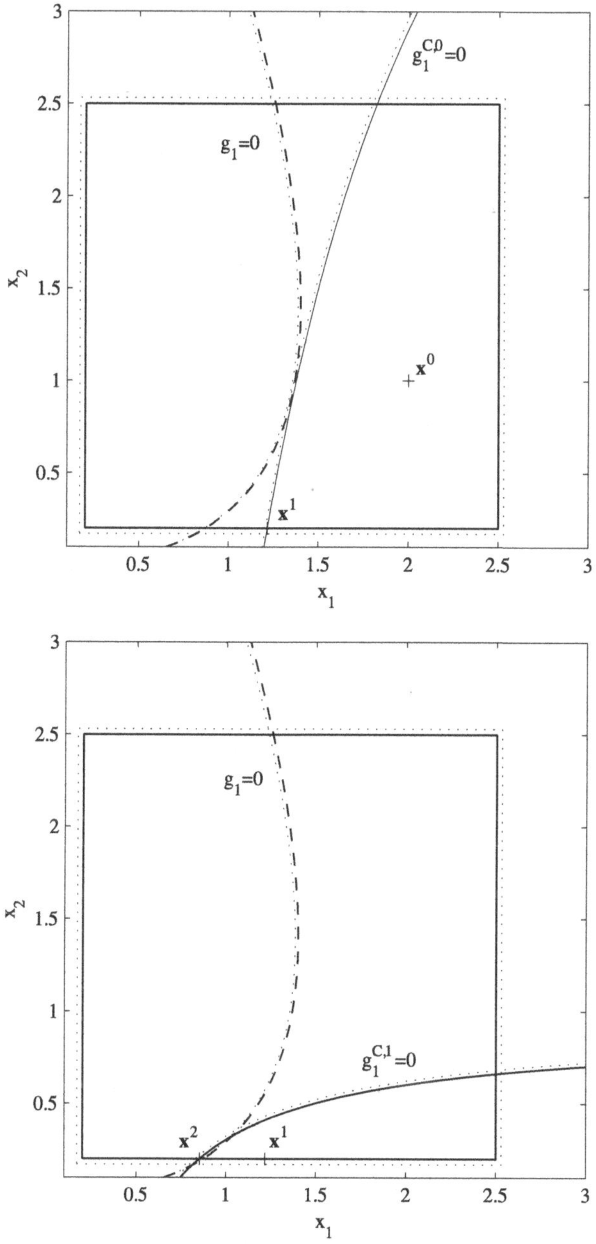

Fig. 4.4 CONLIN approximations of a nonconvex problem

conservative than the function itself. From the figure, we see that $\boldsymbol{x}^2$ is a good approximation of the solution $\boldsymbol{x}^*$ of $(\mathbb{P})_3$. Naturally, we may choose to iterate further in order to get a solution even closer to $\boldsymbol{x}^*$.

4.5 The Method of Moving Asymptotes (MMA)

CONLIN has proven successful for a wide range of structural optimization problems. However, sometimes it converges slowly because of too conservative approximations. On the other hand, sometimes it does not converge at all, indicating that it is not conservative enough. It seems a good idea to have a method where the degree of "conservatism" can be controlled. The Method of Moving Asymptotes (MMA), developed by Svanberg [34], accomplishes just that.

MMA uses the intervening variables

$$y_j(x_j) = \frac{1}{x_j - L_j} \quad \text{or} \quad y_j(x_j) = \frac{1}{U_j - x_j}, \qquad j = 1, \ldots, n,$$

where L_j and U_j are so-called *moving asymptotes* that are changed during the iterations but always satisfy

$$L_j^k < x_j^k < U_j^k, \tag{4.7}$$

for iteration k. The MMA approximation of g_i, $i = 0, \ldots, l$, at the design $\boldsymbol{x}^k$ reads:

$$g_i^{M,k}(\boldsymbol{x}) = r_i^k + \sum_{j=1}^{n} \left(\frac{p_{ij}^k}{U_j^k - x_j} + \frac{q_{ij}^k}{x_j - L_j^k} \right), \tag{4.8}$$

where

$$p_{ij}^k = \begin{cases} (U_j^k - x_j^k)^2 \dfrac{\partial g_i(\boldsymbol{x}^k)}{\partial x_j} & \text{if } \dfrac{\partial g_i(\boldsymbol{x}^k)}{\partial x_j} > 0 \\ 0 & \text{otherwise,} \end{cases} \tag{4.9}$$

$$q_{ij}^k = \begin{cases} 0 & \text{if } \dfrac{\partial g_i(\boldsymbol{x}^k)}{\partial x_j} \geq 0 \\ -(x_j^k - L_j^k)^2 \dfrac{\partial g_i(\boldsymbol{x}^k)}{\partial x_j} & \text{otherwise,} \end{cases} \tag{4.10}$$

$$r_i^k = g_i(\boldsymbol{x}^k) - \sum_{j=1}^{n} \left(\frac{p_{ij}^k}{U_j^k - x_j^k} + \frac{q_{ij}^k}{x_j^k - L_j^k} \right). \tag{4.11}$$

Thus, if p_{ij}^k is not zero, then q_{ij}^k is zero, and vice versa. Differentiation of $g^{M,k}$ twice gives

$$\frac{\partial g_i^{M,k}(\boldsymbol{x})}{\partial x_j} = \frac{p_{ij}^k}{(U_j^k - x_j)^2} - \frac{q_{ij}^k}{(x_j - L_j^k)^2}, \tag{4.12}$$

$$\frac{\partial^2 g_i^{M,k}(\boldsymbol{x})}{\partial x_j^2} = \frac{2p_{ij}^k}{(U_j^k - x_j)^3} + \frac{2q_{ij}^k}{(x_j - L_j^k)^3}, \tag{4.13}$$

$$\frac{\partial^2 g_i^{M,k}(\boldsymbol{x})}{\partial x_j \partial x_p} = 0, \quad \text{if } j \neq p. \tag{4.14}$$

MMA shares some nice features with CONLIN:

- The MMA approximation is a first order approximation, i.e. $g_i^{M,k}(\boldsymbol{x}^k) = g_i(\boldsymbol{x}^k)$ and $\partial g_i^{M,k}(\boldsymbol{x}^k)/\partial x_j = \partial g_i(\boldsymbol{x}^k)/\partial x_j$.
- $g_i^{M,k}$ is an explicit, convex function. The convexity follows since (4.9) and (4.10) imply that $p_{ij}^k \geq 0$ and $q_{ij}^k \geq 0$, and (4.7), (4.13) and (4.14) give that the Hessian $\nabla^2 g_i^{M,k}$ is positive semidefinite.
- The approximation is separable.

The MMA approximation of $(\mathbb{SO})_{\text{nf}}$ at iteration k is written

$$(\text{MMA}) \quad \begin{cases} \min\limits_{\boldsymbol{x}} g_0^{M,k}(\boldsymbol{x}) \\ \text{s.t.} \quad g_i^{M,k}(\boldsymbol{x}) \leq 0, \quad i = 1, \ldots, l \\ \qquad \alpha_j^k \leq x_j \leq \beta_j^k, \quad j = 1, \ldots, n, \end{cases}$$

where α_j^k and β_j^k are move limits to be defined below. This convex, separable problem may be solved using Lagrangian duality. Normally, a term $\varepsilon(U_j^k - x_j^k)^2/(U_j^k - L_j^k)$, where $\varepsilon > 0$, is added to p_{0j}^k and a term $\varepsilon(x_j^k - L_j^k)^2/(U_j^k - L_j^k)$ is added to q_{0j}^k to make the objective function $g_0^{M,k}$ strictly convex.

How do the moving asymptotes affect the MMA approximations? In order to answer this, let us study two sets of moving asymptotes: (L_j^k, U_j^k) and $(\bar{L}_j^k, \bar{U}_j^k)$, where

$$\bar{L}_j^k \leq L_j^k < x_j^k < U_j^k \leq \bar{U}_j^k. \tag{4.15}$$

Introduce the function

$$f_i^{M,k}(\boldsymbol{x}) = g_i^M(\boldsymbol{x}) - \bar{g}_i^{M,k}(\boldsymbol{x}),$$

where $\bar{g}_i^{M,k}(\boldsymbol{x})$ is defined as $g_i^M(\boldsymbol{x})$ although using the moving asymptotes $(\bar{L}_j^k, \bar{U}_j^k)$ instead of (L_j^k, U_j^k). It is easy to show that the only solution to $f_i^{M,k}(\boldsymbol{x}) = 0$ for $L_j^k < x_j < U_j^k$, is $x_j = x_j^k$. Differentiation of $f_i^{M,k}$ gives that $\partial f_i^{M,k}(\boldsymbol{x}^k)/\partial x_j = 0$, whereas the only nonzero elements in the Hessian of $f_i^{M,k}$ at $\boldsymbol{x}^k$ become

$$\frac{\partial^2 f_i^{M,k}(\boldsymbol{x}^k)}{\partial x_j^2} = \begin{cases} \dfrac{2}{U_j^k - x_j^k} g_{i,j} - \dfrac{2}{\bar{U}_j^k - x_j^k} g_{i,j} & \text{if } g_{i,j} \geq 0 \\ -\dfrac{2}{x_j^k - L_j^k} g_{i,j} + \dfrac{2}{x_j^k - \bar{L}_j^k} g_{i,j} & \text{if } g_{i,j} < 0, \end{cases}$$

where $g_{i,j} = \partial g_i(\boldsymbol{x}^k)/\partial x_j$. Because of (4.15), the Hessian is positive semidefinite. Thus, $f_i^{M,k}$ is minimized for $\boldsymbol{x} = \boldsymbol{x}^k$. Since $f_i^{M,k}(\boldsymbol{x}^k) = 0$, we conclude that

$f_i^{M,k}(\boldsymbol{x}) \geq 0$, that is $g_i^{M,k}(\boldsymbol{x}) \geq \bar{g}_i^{M,k}(\boldsymbol{x})$, for $L_j^k < x_j < U_j^k$. This means that if the asymptotes are brought closer to the current design $\boldsymbol{x}^k$ the approximations become larger, i.e. more conservative. By modifying the asymptotes during the iterations we may thus control how conservative the approximations should be. In what follows we describe Svanberg's heuristic approach to update the asymptotes.

For iteration k, the lower asymptote L_j^k and the upper asymptote U_j^k for design variable x_j, $j = 1, \ldots, n$, are updated according to the following rule: for $k = 0$ and $k = 1$,

$$L_j^k = x_j^k - s_{\text{init}}(x_j^{\max} - x_j^{\min}), \tag{4.16}$$

$$U_j^k = x_j^k + s_{\text{init}}(x_j^{\max} - x_j^{\min}), \tag{4.17}$$

where $0 < s_{\text{init}} < 1$, $x_j^{\min}$ and $x_j^{\max}$ are the lower and upper bounds of design variable x_j, and x_j^k is the value of x_j at iteration k. For $k \geq 2$ the signs of $x_j^k - x_j^{k-1}$ and $x_j^{k-1} - x_j^{k-2}$ are studied. If the signs are opposite, the variable x_j oscillates, and therefore the asymptotes L_j^k and U_j^k should be forced closer to x_j^k to make the MMA approximation more conservative. We put

$$L_j^k = x_j^k - s_{\text{slower}}(x_j^{k-1} - L_j^{k-1}),$$

$$U_j^k = x_j^k + s_{\text{slower}}(U_j^{k-1} - x_j^{k-1}),$$

where $0 < s_{\text{slower}} < 1$. If, on the other hand, $x_j^k - x_j^{k-1}$ and $x_j^{k-1} - x_j^{k-2}$ have the same sign, the asymptotes are brought further away from x_j^k in order to make the MMA approximation less conservative and (hopefully) speed up the convergence:

$$L_j^k = x_j^k - s_{\text{faster}}(x_j^{k-1} - L_j^{k-1}),$$

$$U_j^k = x_j^k + s_{\text{faster}}(U_j^{k-1} - x_j^{k-1}),$$

where $s_{\text{faster}} > 1$. In each iteration, the design variables are made to satisfy the constraint

$$\alpha_j^k \leq x_j^k \leq \beta_j^k,$$

where the move limits α_j^k and β_j^k are chosen as

$$\alpha_j^k = \max(x_j^{\min},\ L_j^k + \mu(x_j^k - L_j^k)), \tag{4.18}$$

$$\beta_j^k = \min(x_j^{\max},\ U_j^k - \mu(U_j^k - x_j^k)), \tag{4.19}$$

where $0 < \mu < 1$. It will then always hold that

$$L_j^k < \alpha_j^k \leq x_j^k \leq \beta_j^k < U_j^k,$$

which prevents $U_j^k - x_j^k$ and $x_j^k - L_j^k$ from becoming zero, and thus division by zero is avoided in the MMA approximations.

It turns out that both SLP and CONLIN are special cases of MMA: if $L_j^k = 0$ and $U_j^k \to +\infty$, CONLIN is obtained, and if $L_j^k \to -\infty$ and $U_j^k \to +\infty$, SLP is obtained. To prove the first statement, we first write

$$\frac{1}{U - x_j} = \frac{1}{U(1 - x_j U^{-1})} = U^{-1}(1 + x_j U^{-1} + O(U^{-2})),$$

where $U = U_j^k$, and $O(U^{-2})$ indicates a function that may be written as $U^{-2} f(U)$ as $U \to +\infty$, where $f(U)$ is a bounded function. The MMA approximation becomes

$$\begin{aligned}
g_i^{M,k}(\boldsymbol{x}) &= g_i(\boldsymbol{x}^k) - \sum_+ (U - x_j^k) g_{i,j} \\
&\quad + \sum_- x_j^k g_{i,j} + \sum_+ \frac{(U - x_j^k)^2}{U - x_j} g_{i,j} - \sum_- \frac{(x_j^k)^2}{x_j} g_{i,j} \\
&= g_i(\boldsymbol{x}^k) - \sum_+ (U - x_j^k) g_{i,j} + \sum_- x_j^k g_{i,j} \\
&\quad + \sum_+ \left(U^2 + (x_j^k)^2 - 2U x_j^k\right) U^{-1} \left(1 + x_j U^{-1} + O(U^{-2})\right) g_{i,j} \\
&\quad - \sum_- \frac{(x_j^k)^2}{x_j} g_{i,j} \\
&= g_i(\boldsymbol{x}^k) - \sum_+ (U - x_j^k) g_{i,j} + \sum_- x_j^k g_{i,j} \\
&\quad + \sum_+ \left(U - (x_j^k)^2 U^{-1} - 2x_j^k\right) \left(1 + x_j U^{-1} + O(U^{-2})\right) g_{i,j} \\
&\quad - \sum_- \frac{(x_j^k)^2}{x_j} g_{i,j} \\
&= g_i(\boldsymbol{x}^k) + \sum_+ x_j^k g_{i,j} + \sum_- x_j^k g_{i,j} \\
&\quad + \sum_+ \left(x_j + O(U^{-1}) - 2x_j^k\right) g_{i,j} - \sum_- \frac{(x_j^k)^2}{x_j} g_{i,j},
\end{aligned}$$

where $g_{i,j} = \partial g_i(\boldsymbol{x}^k)/\partial x_j$ and $\sum_+$ indicates summation over terms where $g_{i,j} > 0$, and similarly for $\sum_-$. Letting $U \to +\infty$ in this expression, we obtain

$$g_i^{M,k}(\boldsymbol{x}) \to g_i(\boldsymbol{x}^k) + \sum_{+}(x_j - x_j^k)g_{i,j} + \sum_{-}\left(x_j^k - \frac{(x_j^k)^2}{x_j}\right)g_{i,j},$$

which coincides with the definition of the CONLIN approximation in (4.5).

Example 4.3 Consider the same function g as in Example 4.1. We would like to calculate the MMA approximation of this function at $x^0 = 1$. Since the derivative $g_x(x^0) = 2.9 > 0$, g is linearized in the variable $1/(U^0 - x)$. In Fig. 4.5 the approximations for different values of the upper asymptote are shown. Note how the approximation becomes less conservative as U^0 is brought further away from x^0. For $U^0 = 10^4$, the approximation is almost linear, which is in agreement with the statement above that SLP is obtained as $U \to +\infty$ (and $L \to -\infty$).

Example 4.4 The constraint function g_1 in Exercise 4.2 is to be approximated at $\boldsymbol{x}^0 = (2, 1)$ using MMA. We have that $\partial g_1(\boldsymbol{x}^0)/\partial x_1 = -0.041$ and $\partial g_1(\boldsymbol{x}^0)/\partial x_2 = 0.019$, so that g_1 is linearized in $1/(x_1 - L_1^0)$ and $1/(U_2^0 - x_2)$. The asymptotes are calculated according to (4.16) and (4.17). First $s_{\text{init}} = 0.2$, which gives $L_1^0 = 1.54$ and $U_2^0 = 1.46$. The asymptotes are moved further away from $\boldsymbol{x}^0$

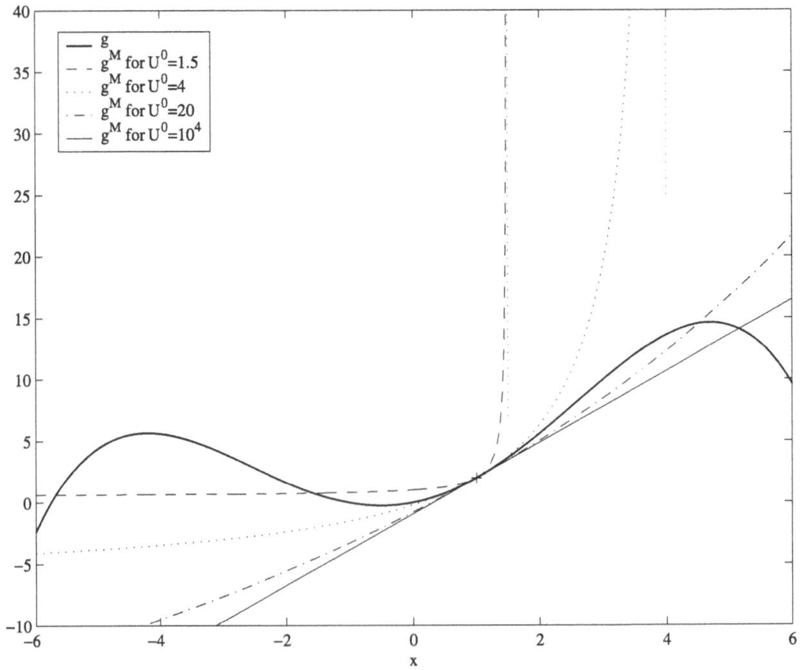

Fig. 4.5 MMA approximations. The two *dotted vertical lines* represent the asymptotes of g^M for the case $U^0 = 1.5$ and $U^0 = 4$, respectively

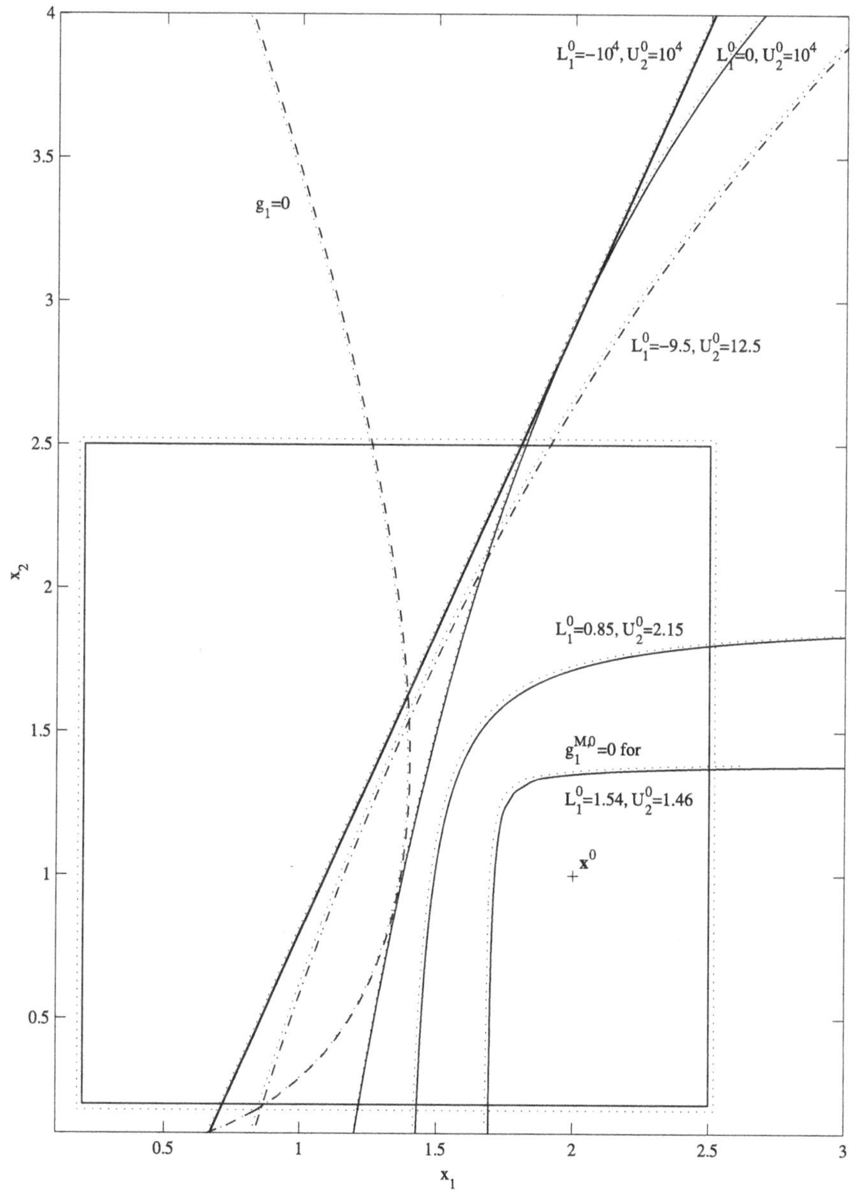

Fig. 4.6 MMA approximations of a nonconvex problem

by increasing s_{init} to 0.5 and 5.0 resulting in $L_1^0 = 0.85$, $U_2^0 = 2.15$ and $L_1^0 = -9.5$, $U_2^0 = 12.5$, respectively. This should make the approximations less conservative, which can also be verified in Fig. 4.6. CONLIN is simulated by putting $L_1^0 = 0$ and $U_2^0 = 10^4$, whereas SLP is simulated by letting $L_1^0 = -10^4$ and $U_2^0 = 10^4$.

A CONLIN approximation is always more, or equally, conservative than an SLP approximation, which is also seen in the figure.

4.6 Exercises

Exercise 4.1 Is the CONLIN approximation of a linear function always exact (i.e. the function itself)? What is the case for MMA?

Exercise 4.2 Prove that CONLIN and MMA are first order approximations.

Exercise 4.3 Show that if the asymptotes are chosen as $L_j^k \to -\infty$ and $U_j^k \to +\infty$ in MMA, then the SLP approximation is obtained.

Exercise 4.4 One wants to maximize the stiffness of the truss in Exercise 3.4 by instead minimizing the size of the displacement vector, or $\boldsymbol{u}^T\boldsymbol{u}$. Defining $x_i = lA_i/V_0,\ i = 1, \ldots, 3$, this leads to the following optimization problem:

$$\min \frac{1}{x_1^2} + \frac{1}{x_2^2} + \frac{4}{x_3^2} + \frac{1}{x_1x_2} + \frac{2\sqrt{2}}{x_1x_3} + \frac{2\sqrt{2}}{x_2x_3}$$

$$\text{s.t.} \quad x_1 + x_2 + \sqrt{2}x_3 - 1 \leq 0, \qquad x_1, x_2, x_3 \geq 0.$$

a) Show that the problem is convex. Note that in Sect. 2.6, we studied a problem where $\boldsymbol{u}^T\boldsymbol{u}$ turned out to be a nonconvex function of the design variables.

b) Obtain a subproblem by performing a CONLIN approximation of the problem at $x_i = 1,\ i = 1, \ldots, 3$. Solve the subproblem.

Exercise 4.5 The four-bar truss in Fig. 4.7 is subjected to the force $P > 0$. The cross-sectional areas A_1, A_2, A_3, A_4 of the truss should be determined such that the displacement $u_x^{\mathcal{B}}$ is minimized. The volume of the truss is not allowed to exceed the value V_0.

a) Formulate the problem as a mathematical programming problem. Introduce new design variables as $x_i = lA_i/V_0,\ i = 1, \ldots, 4$.

b) What is the optimum cross-sectional area A_1^* of bar 1?

c) Obtain the CONLIN approximation of the problem at (the nonfeasible point) $x_i = 1,\ i = 1, \ldots, 4$. Solve the CONLIN approximation of the problem by using Lagrangian duality.

Exercise 4.6 In this exercise you are going to use a MATLAB program to solve some simple sizing optimization problems using MMA. The aim of the exercise is to get a better understanding of how MMA works by visualizing the approximated subproblems generated by the algorithm. The code is available from the book's homepage (www.mechanics.iei.liu.se/edu_ug/strop/).

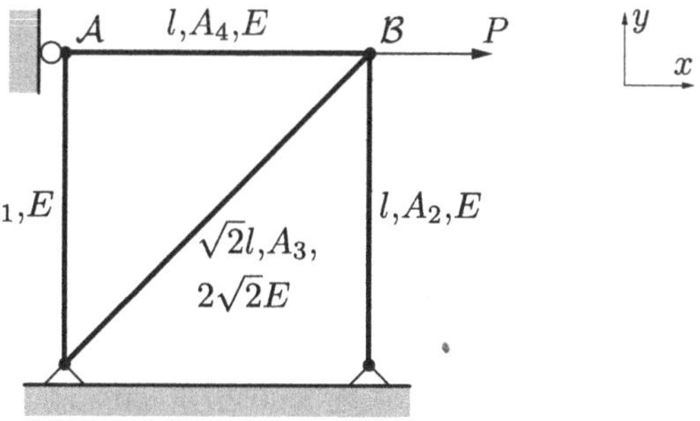

Fig. 4.7 The four-bar truss of Exercise 4.5

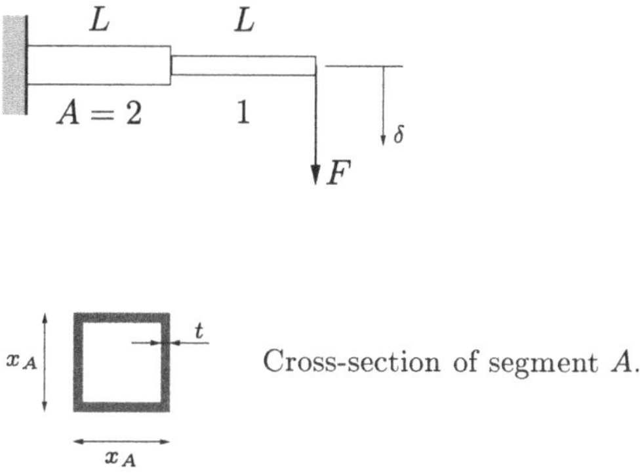

Fig. 4.8 The two-segment cantilever of Exercise 4.6

Consider the problem of minimizing the weight of a beam consisting of two segments, see Fig. 4.8, and cf. Sect. 2.4. The problem may be written as a mathematical programming problem as

$$
(\mathbb{P})_4 \quad \begin{cases} \min\limits_{x_1,x_2} g_0(x_1,x_2) = x_1 + x_2 \\ \text{s.t.} \quad \begin{cases} g_1(x_1,x_2) = \dfrac{1}{x_1^3} + \dfrac{7}{x_2^3} - 1 \leq 0 \\ 0.1 \leq x_1 \leq 10, \qquad 0.1 \leq x_2 \leq 10, \end{cases} \end{cases}
$$

where we have introduced other lower and upper bounds on the design variables than previously. We will solve this problem by using MMA. The asymptotes are updated during the iterations according to the rules described on page 68 ff.

The script `main_cantilever.m` is the main program file to solve $(\mathbb{P})_1$ using MMA. In it, the number of design variables, constraints, lower and upper bounds of the design variables, the starting point, the maximum number of iterations and MMA parameters are given. The objective function, the constraint function and their derivatives are supplied in the function file `cantilever_2.m`. Run the program by typing `main_cantilever` at the MATLAB prompt. At each iteration, the current solution and the MMA approximation of the constraint g_1 are plotted. The contour lines of the exact objective function are also plotted. The starting point has been put to $(x_1, x_2) = (5, 5)$ and $s_{\text{init}} = 0.1$. As you can see the iterates converge to the point obtained analytically in Sect. 2.4.

1. Now make the MMA approximation less conservative by putting $s_{\text{init}} = 0.3$. What happens? Keep $s_{\text{init}} = 0.3$, but make MMA more conservative by changing s_{slower} from 0.7 to 0.5, and s_{faster} from 1.2 to 1.1. What happens? Run the script `plottie`. In MATLAB Fig. 2, the iteration histories of x_1, L_1 and U_1, and α_1 and β_1 are plotted (you may click and drag the legend box if it obstructs the curves). Fig. 3 shows the same for the second design variable, whereas Fig. 4 shows the iteration histories of g_0 and g_1. Keep $s_{\text{slower}} = 0.5$ and $s_{\text{faster}} = 1.1$, but put $s_{\text{init}} = 0.1$ again. Run `plottie` and note the difference in appearance of the plots.
2. Outcomment the call of the function `mma_solver`, and instead call `mma_solver_mod`. This MMA version updates the asymptotes and move limits in a different way:

$$
\begin{aligned}
L_j^k &= t\, x_j^k, \\
U_j^k &= x_j^k / t, \\
\alpha_j^k &= \max(0.5 x_j^k,\ 1.01 L_j^k,\ x_j^{\min}), \\
\beta_j^k &= \min(2 x_j^k,\ 0.99 U_j^k,\ x_j^{\max}),
\end{aligned}
$$

where $t > 0$. In the code, $t = 10^{-10}$, which means that a CONLIN approximation is obtained (why?). Does CONLIN manage to solve the problem? Run `plottie_conlin`, which works in the same way as `plottie`, except that the upper bounds U_j^k are not plotted as these will be very large.
3. For a general optimization problem it may be very difficult to find a starting point that is feasible, i.e. a point satisfying the constraints $g_i(\mathbf{x}) \leq 0,\ i = 1, \ldots, l$. Therefore, it is important that solution algorithms are able to find (local) optima even if started outside the feasible set. Use CONLIN again, but modify the starting point to $(x_1, x_2) = (1, 1)$, which is outside the feasible set. What happens? Switch back to the "ordinary" MMA solver again (`mma_solver`) and use $s_{\text{init}} = 0.1$, $s_{\text{slower}} = 0.7$ and $s_{\text{faster}} = 1.2$. What happens?
4. The volume of the four-bar truss in Example 4.2 should be minimized under the constraint that the elongation of bar 1, $\delta \leq \delta_0$. Derive the nonconvex optimization problem $(\mathbb{P})_3$ on page 64. We will solve it numerically by generating a sequence of convex subproblems.
5. The script `main_four_bar.m` is used to solve $(\mathbb{P})_2$. All data of the objective and constraint functions are in the file `four_bar.m`. Solve the problem by using the starting point $(x_1, x_2) = (2, 1.5)$, and the MMA parameters $s_{\text{init}} = 0.5$, $s_{\text{slower}} = 0.7$ and $s_{\text{faster}} = 1.2$. What is the solution? What is the optimum objective value? Also try to use the starting point $(x_1, x_2) = (1.3, 2.4)$. Note: when using the latter starting point, you should outcomment the line "`y1=x_min(2);`" in `main_four_bar.m` in order to get a nice plot.
6. Put $s_{\text{init}} = 10^{12}$, which will simulate Sequential Linear Programming (SLP) (why?). Use the two starting points in task 5. Does it work?

7. Solve the problem by CONLIN using the two starting points in task 5. What happens?

8. Let us assume that x_1 is *fixed* at $x_1 = 1.35$, and that only x_2 may be varied in the optimization. For this case, the script `main_four_bar_2.m` is used. All data for the objective and constraint functions are in the file `four_bar_2.m`. Use the same MMA parameters as in task 5. Choose a number of different starting points: $x_2 = 2.45, 1.5, 1.4, 1.3, 0.5$. What happens? The behavior is typical for nonconvex problems where there exist several local minima. You can plot the iteration history by running `plottie_2`. What is the global minimum of this problem? What are the optimal objective values obtained by MMA? Are they smaller or greater than the one obtained in task 5? Why?

Chapter 5
Sizing Stiffness Optimization of a Truss

In this chapter we will describe in detail how sequential explicit approximations can be used to solve a particular large-scale structural optimization problem, namely that of determining the cross-sectional areas of the bars in a two-dimensional truss with fixed locations of the nodes so that its stiffness is maximized.

5.1 The Simultaneous Formulation of the Problem

In order to maximize the stiffness of a truss, see Fig. 5.1, we naturally need to introduce a suitable measure of stiffness. Here, we will choose to use the compliance C of the truss, i.e. $\boldsymbol{F}^T\boldsymbol{u}$, where $\boldsymbol{u}$ are the displacements of the nonsuppressed nodes of the truss, and $\boldsymbol{F}$ are the given external forces at these nodes. It should be clear that if the compliance is small, the truss will be stiff.

One can easily conceive other measures of stiffness, such as the size of the displacement vector, or $\boldsymbol{u}^T\boldsymbol{u}$. Compliance is a much more popular measure, however. There are at least two reasons for this. First, in a nested formulation, the compliance is a convex function of the design variables, i.e. the cross-sectional areas of the bars, whereas as we have seen in Sect. 2.6, $\boldsymbol{u}^T\boldsymbol{u}$ can be a nonconvex function of these variables. Second, in a truss where the compliance has been minimized for a given amount of material, all bars have the same stress. Intuitively, one has the impression that good use of the available material has been made if all bars have the same stress. Precise formulations and proofs of these statements will be given later.

The optimization problem we are faced with then, may be written as follows in a simultaneous formulation

$$(\mathbb{P})_{\text{sf}} \quad \begin{cases} \min\limits_{\boldsymbol{x},\boldsymbol{u}} \boldsymbol{F}^T\boldsymbol{u} \\ \text{s.t.} \quad \boldsymbol{K}(\boldsymbol{x})\boldsymbol{u} = \boldsymbol{F} \\ \qquad \sum\limits_{j=1}^{n} l_j x_j \le V_{\max} \\ \qquad \boldsymbol{x} \in \mathcal{X} = \{\boldsymbol{x} \in \mathbb{R}^n : x_j^{\min} \le x_j \le x_j^{\max},\ j = 1, \dots, n\}, \end{cases}$$

where n is the number of bars, l_j is the length of bar j, x_j is the cross-sectional area of bar j, and $V_{\max}$ is the maximum allowed volume of the truss. For simplicity, we have assumed that the global external force vector $\boldsymbol{F}$ does not depend on the design. In general, $\boldsymbol{F}$ may very well depend on $\boldsymbol{x}$, e.g. by including the weight of the bars. The matrix $\boldsymbol{K}(\boldsymbol{x})$ is the global stiffness matrix of the structure, and there

P.W. Christensen, A. Klarbring, *An Introduction to Structural Optimization*,

Fig. 5.1 A truss to be optimized

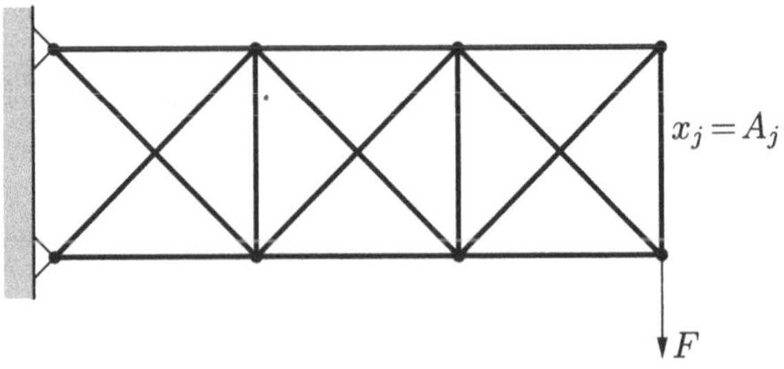

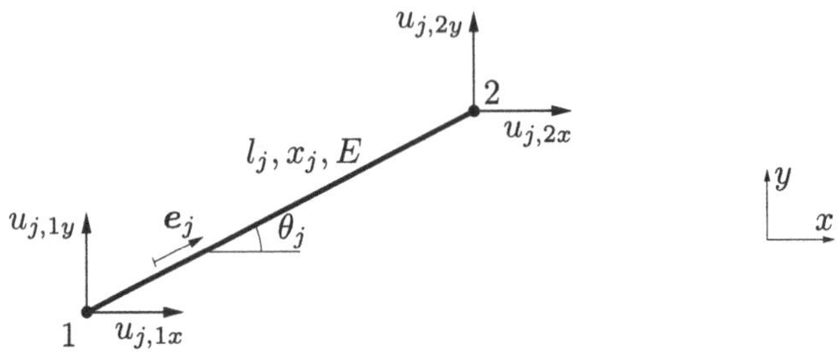

Fig. 5.2 A general bar j in the truss

are given lower and upper bounds $x_j^{\min}$ and $x_j^{\max}$ on the design variables. It holds that $x_j^{\min} \geq 0$ and $x_j^{\max}$ is finite. Next, the matrix $\boldsymbol{K}(\boldsymbol{x})$ will be derived. To that end, study a general bar j, and let local node numbers 1 and 2 denote the end points of the bar, cf. Fig. 5.2. A unit vector $\boldsymbol{e}_j$ along the bar is defined so that it points from node 1 to node 2. The orientation of the bar is determined by the angle θ_j, which is the angle from the x-axis to $\boldsymbol{e}_j$, measured anti-clockwise, i.e. around the z-axis. Thus, $\boldsymbol{e}_j$ may be written

$$\boldsymbol{e}_j = \begin{bmatrix} \cos\theta_j \\ \sin\theta_j \end{bmatrix}.$$

The displacements of the end points of bar j are collected in a vector

$$\boldsymbol{u}_j = \begin{bmatrix} \boldsymbol{u}_{j,1} \\ \boldsymbol{u}_{j,2} \end{bmatrix}, \quad \text{where} \quad \boldsymbol{u}_{j,1} = \begin{bmatrix} u_{j,1x} \\ u_{j,1y} \end{bmatrix} \quad \text{and} \quad \boldsymbol{u}_{j,2} = \begin{bmatrix} u_{j,2x} \\ u_{j,2y} \end{bmatrix}.$$

The elongation δ_j of bar j is $(\boldsymbol{u}_{j,2} - \boldsymbol{u}_{j,1}) \cdot \boldsymbol{e}_j$, or

$$\delta_j = \boldsymbol{B}_j \boldsymbol{u}_j, \tag{5.1}$$

where

$$\boldsymbol{B}_j = [\,-\boldsymbol{e}_j^T \quad \boldsymbol{e}_j^T\,]. \tag{5.2}$$

The external force $\boldsymbol{f}_j$ on the end points of bar j may be written

$$\boldsymbol{f}_j = \boldsymbol{B}_j^T s_j, \tag{5.3}$$

where s_j is the force in the bar. If $s_j > 0$, bar j is in tension, otherwise it is in compression. From Hooke's law we obtain the relation between the bar force and the elongation as

$$s_j = \sigma_j x_j = E\varepsilon_j x_j = \frac{E\delta_j x_j}{l_j} = D_j \delta_j,$$

where σ_j is the stress in the bar, ε_j is the strain, E is Young's modulus, which is assumed to be the same for all bars in the truss, and

$$D_j = \frac{E x_j}{l_j}. \tag{5.4}$$

If this is inserted into (5.3), we get, upon using (5.1), that

$$\boldsymbol{f}_j = \boldsymbol{k}_j \boldsymbol{u}_j, \tag{5.5}$$

where

$$\boldsymbol{k}_j = \boldsymbol{B}_j^T D_j \boldsymbol{B}_j \tag{5.6}$$

is the element stiffness matrix of bar j. It will prove useful to write $\boldsymbol{k}_j = \boldsymbol{k}_j(\boldsymbol{x})$ on the form

$$\boldsymbol{k}_j(\boldsymbol{x}) = x_j \boldsymbol{k}_j^0, \tag{5.7}$$

where the constant matrix $\boldsymbol{k}_j^0$ is obtained from (5.6) as

$$\boldsymbol{k}_j^0 = \frac{E}{l_j}\begin{bmatrix} c^2 & sc & -c^2 & -sc \\ sc & s^2 & -sc & -s^2 \\ -c^2 & -sc & c^2 & sc \\ -sc & -s^2 & sc & s^2 \end{bmatrix}, \tag{5.8}$$

where $s = \sin\theta_j$ and $c = \cos\theta_j$. The matrix $\boldsymbol{k}_j^0$ represents the element stiffness matrix for bar j per unit area.

The element displacement vector $\boldsymbol{u}_j$ for bar j may be obtained from the global displacement vector $\boldsymbol{u}$ as

$$\boldsymbol{u}_j = \boldsymbol{C}_j \boldsymbol{u}, \tag{5.9}$$

where $\boldsymbol{C}_j$ is a matrix with elements 0 and 1. It should be noted that $\boldsymbol{u}_j$ holds all of the displacements of the end points of bar j, including those that are always zero due to supports. In $\boldsymbol{u}$, however, displacements that are always zero are not included. This means that there is at most one element 1 for each row in $\boldsymbol{C}_j$. By multiplying (5.5) by $\boldsymbol{C}_j^T$ and summing over all bars, we get the global equilibrium equations for the truss as

$$\boldsymbol{F} = \boldsymbol{K}(\boldsymbol{x})\boldsymbol{u}, \tag{5.10}$$

where

$$K(x) = \sum_{j=1}^{n} K_j(x), \quad K_j(x) = C_j^T k_j(x) C_j. \tag{5.11}$$

Here, $K(x)$ is the global stiffness matrix of the truss. The matrix $K_j(x)$ is a global version of the element stiffness matrix $k_j(x)$ whose nonzero elements are the elements of $k_j(x)$ that correspond to the degrees-of-freedom in the global displacement vector u. In Example 5.1 below, the calculation of $K(x)$ will be illustrated for a small truss.

The matrix $K(x)$ may also be expressed as

$$K(x) = \sum_{j=1}^{n} x_j K_j^0, \quad K_j^0 = C_j^T k_j^0 C_j, \tag{5.12}$$

where K_j^0 is a constant matrix. Transposing $K_j(x)$, we obtain $C_j^T (C_j^T k_j(x))^T = C_j^T k_j(x) C_j$, since $k_j(x)$ is symmetric; cf. (5.8). Thus, $K(x)$ is symmetric. Also, in (5.10),

$$F = \sum_{j=1}^{n} C_j^T f_j. \tag{5.13}$$

In this sum, the contribution from the unknown reaction forces from supports and the likewise unknown forces from neighboring bars, will be zero. Consequently, we may rewrite (5.13) as

$$F = \sum_{j=1}^{n} C_j^T f_j^a, \tag{5.14}$$

where f_j^a is the vector of the given applied forces on the end points of bar j. Thus, the vector F is the total applied force on the truss. The total element external forces f_j may be calculated from (5.5) once (5.10) has been solved for the displacements.

The process of calculating the global stiffness matrix and the global applied force vector from their counterparts on the element level is called *assembly*. In an actual implementation of this, the matrices C_j are never formed; instead one simply keeps track of on which rows and columns of the global stiffness matrix a certain element stiffness matrix should be added, and similarly for the force vectors. One often simply writes

$$K(x) = \mathop{\mathbf{A}}_{j=1}^{n} k_j(x), \qquad F = \mathop{\mathbf{A}}_{j=1}^{n} f_j^a \tag{5.15}$$

to describe this process.

From (5.1) and (5.9), the elongations δ of all bars in the truss are obtained as

$$\delta = \bar{B} u,$$

where

$$\bar{\boldsymbol{B}} = \begin{bmatrix} \boldsymbol{B}_1\boldsymbol{C}_1 \\ \vdots \\ \boldsymbol{B}_n\boldsymbol{C}_n \end{bmatrix}.$$

If (5.3) is written for the whole truss, it becomes $\boldsymbol{F} = \bar{\boldsymbol{B}}^T \boldsymbol{s}$, where $\boldsymbol{s}$ are the bar forces in all bars. For a statically determinate truss, $\bar{\boldsymbol{B}}$ is invertible, and one obtains

$$\boldsymbol{u} = \bar{\boldsymbol{B}}^{-1}\boldsymbol{\delta} = \bar{\boldsymbol{B}}^{-1} \operatorname{diag}\left(\frac{l_1}{Ex_1}, \ldots, \frac{l_n}{Ex_n}\right) \bar{\boldsymbol{B}}^{-T} \boldsymbol{F},$$

which shows that the displacements vary as $1/x_j$, as claimed in Sect. 4.4. Here, $\operatorname{diag}(A_{11}, \ldots, A_{nn})$ denotes a diagonal matrix with the diagonal elements $A_{11}, \ldots, A_{nn}$. Similarly, the stresses $\boldsymbol{\sigma}$ of all bars may be written

$$\boldsymbol{\sigma} = \operatorname{diag}\left(\frac{1}{x_1}, \ldots, \frac{1}{x_n}\right) \bar{\boldsymbol{B}}^{-T} \boldsymbol{F},$$

and, thus, the stresses also vary as $1/x_j$.

For later use, the expression for the strain energy of bar j will be derived. It is defined as

$$\begin{aligned} U_j &= \frac{1}{2}\int \sigma_j \varepsilon_j \, dV_j = \frac{1}{2} E\varepsilon_j^2 x_j l_j = \frac{1}{2} E \frac{(\boldsymbol{B}_j\boldsymbol{u}_j)^2}{l_j^2} x_j l_j \\ &= \frac{1}{2}(\boldsymbol{B}_j\boldsymbol{u}_j)^T \frac{Ex_j}{l_j} \boldsymbol{B}_j\boldsymbol{u}_j = \frac{1}{2}\boldsymbol{u}_j^T \boldsymbol{B}_j^T D_j \boldsymbol{B}_j \boldsymbol{u}_j, \end{aligned} \tag{5.16}$$

for any displacements $\boldsymbol{u}_j$. From (5.6), we then get that

$$U_j = \frac{1}{2}\boldsymbol{u}_j^T \boldsymbol{k}_j \boldsymbol{u}_j. \tag{5.17}$$

From the last expression in the first row of (5.16), the strain energy is clearly nonnegative, so the stiffness matrix is positive semidefinite. The strain energy U of the whole truss is obtained by summation over all bars:

$$\begin{aligned} U &= \sum_{j=1}^{n} U_j = \frac{1}{2}\sum_{j=1}^{n} (\boldsymbol{C}_j\boldsymbol{u})^T \boldsymbol{k}_j (\boldsymbol{C}_j\boldsymbol{u}) \\ &= \frac{1}{2}\boldsymbol{u}^T \left(\sum_{j=1}^{n} \boldsymbol{C}_j^T \boldsymbol{k}_j \boldsymbol{C}_j\right) \boldsymbol{u} = \frac{1}{2}\boldsymbol{u}^T \boldsymbol{K}\boldsymbol{u}, \end{aligned} \tag{5.18}$$

where we have made use of (5.9) and (5.11). If all bars have a strictly positive cross-sectional area, and the truss is properly anchored, so that there does not exist any nonzero displacement $\boldsymbol{u}$ with zero strain energy of the truss, i.e. a rigid body displacement, $\boldsymbol{K}$ is positive definite, and, hence, invertible.

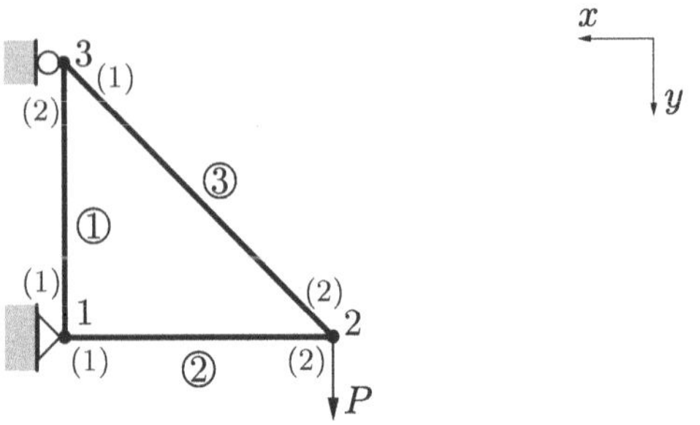

Fig. 5.3 The three-bar truss of Exercise 3.4

Example 5.1 Let us study the three-bar truss of Exercise 3.4. We wish to write down the equilibrium equations (5.10) for the truss by using the approach described in this section. Local and global node numbers are defined as in Fig. 5.3. The displacements of the bars are $\boldsymbol{u}_1 = [0\ 0\ 0\ u_{3y}]^T$, $\boldsymbol{u}_2 = [0\ 0\ u_{2x}\ u_{2y}]^T$, $\boldsymbol{u}_3 = [0\ u_{3y}\ u_{2x}\ u_{2y}]^T$, and the global displacement vector is $\boldsymbol{u} = [u_{2x}\ u_{2y}\ u_{3y}]^T$. By writing $\boldsymbol{u}_j = \boldsymbol{C}_j\boldsymbol{u}$, for $j = 1, 2, 3$, we may identify the matrices $\boldsymbol{C}_j$ as

$$\boldsymbol{C}_1 = \begin{bmatrix} 0 & 0 & 0 \\ 0 & 0 & 0 \\ 0 & 0 & 0 \\ 0 & 0 & 1 \end{bmatrix}, \qquad \boldsymbol{C}_2 = \begin{bmatrix} 0 & 0 & 0 \\ 0 & 0 & 0 \\ 1 & 0 & 0 \\ 0 & 1 & 0 \end{bmatrix}, \qquad \boldsymbol{C}_3 = \begin{bmatrix} 0 & 0 & 0 \\ 0 & 0 & 1 \\ 1 & 0 & 0 \\ 0 & 1 & 0 \end{bmatrix}.$$

For bar 1, $\theta_1 = 3\pi/2$, so $\boldsymbol{e}_1 = [0\ -1]^T$ and $\boldsymbol{B}_1 = [0\ 1\ 0\ -1]$ which gives

$$\boldsymbol{k}_1 = \frac{Ex_1}{l}\begin{bmatrix} 0 & 0 & 0 & 0 \\ 0 & 1 & 0 & -1 \\ 0 & 0 & 0 & 0 \\ 0 & -1 & 0 & 1 \end{bmatrix},$$

according to (5.6). For bar 2, $\theta_2 = \pi$, which gives $\boldsymbol{e}_2 = [-1\ 0]^T$, $\boldsymbol{B}_2 = [1\ 0\ -1\ 0]$ and

$$\boldsymbol{k}_2 = \frac{Ex_2}{l}\begin{bmatrix} 1 & 0 & -1 & 0 \\ 0 & 0 & 0 & 0 \\ -1 & 0 & 1 & 0 \\ 0 & 0 & 0 & 0 \end{bmatrix}.$$

Finally, for bar 3, $\theta_3 = 3\pi/4$, which results in $\boldsymbol{e}_3 = [-1\ \ 1]^T/\sqrt{2}$, $\boldsymbol{B}_3 = [1\ -1\ -1\ 1]/\sqrt{2}$ and

$$\boldsymbol{k}_3 = \frac{Ex_3}{2\sqrt{2}l}\begin{bmatrix} 1 & -1 & -1 & 1 \\ -1 & 1 & 1 & -1 \\ -1 & 1 & 1 & -1 \\ 1 & -1 & -1 & 1 \end{bmatrix}.$$

The global versions $\boldsymbol{K}_1$, $\boldsymbol{K}_2$ and $\boldsymbol{K}_3$ of the element stiffness matrices $\boldsymbol{k}_1$, $\boldsymbol{k}_2$ and $\boldsymbol{k}_3$ are obtained from (5.11) as

$$\boldsymbol{K}_1 = \frac{Ex_1}{l}\begin{bmatrix} 0 & 0 & 0 \\ 0 & 0 & 0 \\ 0 & 0 & 1 \end{bmatrix}, \qquad \boldsymbol{K}_2 = \frac{Ex_2}{l}\begin{bmatrix} 1 & 0 & 0 \\ 0 & 0 & 0 \\ 0 & 0 & 0 \end{bmatrix},$$

$$\boldsymbol{K}_3 = \frac{Ex_3}{2\sqrt{2}l}\begin{bmatrix} 1 & -1 & 1 \\ -1 & 1 & -1 \\ 1 & -1 & 1 \end{bmatrix}.$$

The global stiffness matrix then becomes

$$\boldsymbol{K} = \sum_{j=1}^{3} \boldsymbol{K}_j = \frac{E}{l}\begin{bmatrix} x_2 + \frac{x_3}{2\sqrt{2}} & -\frac{x_3}{2\sqrt{2}} & \frac{x_3}{2\sqrt{2}} \\ -\frac{x_3}{2\sqrt{2}} & \frac{x_3}{2\sqrt{2}} & -\frac{x_3}{2\sqrt{2}} \\ \frac{x_3}{2\sqrt{2}} & -\frac{x_3}{2\sqrt{2}} & x_1 + \frac{x_3}{2\sqrt{2}} \end{bmatrix}.$$

The applied force vectors of the bars are $\boldsymbol{f}_1^a = [0\ 0\ 0\ 0]^T$, $\boldsymbol{f}_2^a = [0\ 0\ 0\ 0]^T$, $\boldsymbol{f}_3^a = [0\ 0\ 0\ P]^T$. It is immaterial whether the force P is chosen to act on bar 2 or 3. The global applied force vector of (5.14) becomes

$$\boldsymbol{F} = \sum_{j=1}^{3} \boldsymbol{C}_j^T \boldsymbol{f}_j^a = \begin{bmatrix} 0 \\ P \\ 0 \end{bmatrix}.$$

Solution of $\boldsymbol{K}\boldsymbol{u} = \boldsymbol{F}$ gives

$$\boldsymbol{u} = \begin{bmatrix} u_{2x} \\ u_{2y} \\ u_{3y} \end{bmatrix} = \frac{Pl}{E}\begin{bmatrix} \frac{1}{x_2} \\ \frac{1}{x_1} + \frac{1}{x_2} + \frac{2\sqrt{2}}{x_3} \\ \frac{1}{x_1} \end{bmatrix}.$$

Finally, we calculate the element external force vectors as $\boldsymbol{f}_j = \boldsymbol{k}_j \boldsymbol{u}_j$, $j = 1, 2, 3$. The results are $\boldsymbol{f}_1 = P[0\ -1\ 0\ 1]^T$, $\boldsymbol{f}_2 = P[-1\ 0\ 1\ 0]^T$, $\boldsymbol{f}_3 = P[1\ -1\ -1\ 1]^T$, see Fig. 5.4. It should be clear from the figure that these forces keep the truss in equilibrium. It is also readily checked that

$$\boldsymbol{F} = \sum_{j=1}^{n} \boldsymbol{C}_j^T \boldsymbol{f}_j^a = \sum_{j=1}^{n} \boldsymbol{C}_j^T \boldsymbol{f}_j.$$

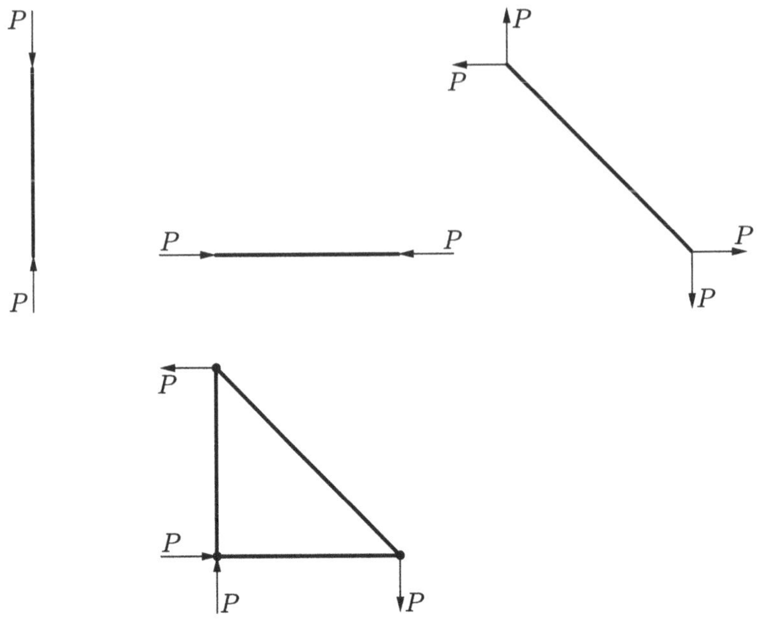

Fig. 5.4 The forces on each bar and on the whole truss

5.2 The Nested Formulation and Some of Its Properties

If the global stiffness matrix is nonsingular we may eliminate the displacement vector $\boldsymbol{u}$ from the simultaneous formulation in order to get the nested formulation

$$(\mathbb{P})_{\mathrm{nf}} \quad \begin{cases} \min\limits_{\boldsymbol{x}} \boldsymbol{F}^T \boldsymbol{u}(\boldsymbol{x}) \\ \text{s.t.} \quad \sum\limits_{j=1}^{n} l_j x_j - V_{\max} \leq 0 \\ \qquad \boldsymbol{x} \in \mathcal{X}, \end{cases}$$

where $\boldsymbol{x} \mapsto \boldsymbol{u}(\boldsymbol{x})$ is an implicit function defined through the equilibrium equations $\boldsymbol{K}(\boldsymbol{x})\boldsymbol{u}(\boldsymbol{x}) = \boldsymbol{F}$. It will be assumed that there is a feasible point, and consequently, since the feasible set is compact, a solution, to $(\mathbb{P})_{\mathrm{nf}}$. That is, we assume that the lower bounds on $\boldsymbol{x}$ are not so large that the volume of the truss is always greater than $V_{\max}$.

If $x_j^{\min} = 0,\ j = 1, \ldots, n$, bars may disappear from the ground structure that one starts with, i.e. the sizing optimization problem becomes a topology optimization problem. In this case, the global stiffness matrix will typically be singular for the optimum truss. In Fig. 5.5, a ground structure and three possible optimum solutions are depicted. For case a), $\boldsymbol{K}(\boldsymbol{x})$ is positive definite, whereas for case b), $\boldsymbol{K}(\boldsymbol{x})$ is singular since the rows and columns in $\boldsymbol{K}(\boldsymbol{x})$ corresponding to displacements of the upper-right node will be zero. Similarly, $\boldsymbol{K}(\boldsymbol{x})$ is singular for case c) as well. If $\boldsymbol{K}(\boldsymbol{x})$ is singular, we cannot formulate the nested formulation, but have to solve the much larger simultaneous formulation instead. The reader is referred to Achtziger [1] for a

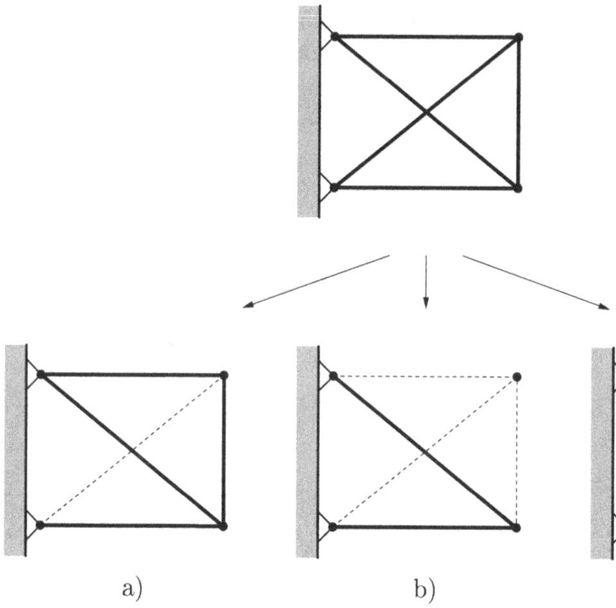

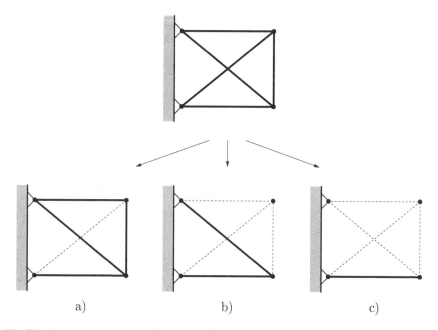

Fig. 5.5 A ground structure and three optimal structures. *Dashed lines* correspond to bars with zero cross-sectional area

solution of the problem using the simultaneous formulation $(\mathbb{P})_{\mathrm{sf}}$. We may avoid the simultaneous formulation, however, by putting the lower bounds in $(\mathbb{P})_{\mathrm{nf}}$ to a small, positive value: $x_j^{\min} = \varepsilon > 0$. It may be proven that as $\varepsilon \to 0$, the solution of $(\mathbb{P})_{\mathrm{nf}}$ for $x_j^{\min} = \varepsilon$ approaches that of $(\mathbb{P})_{\mathrm{sf}}$ for $x_j^{\min} = 0$. When using this technique, we delete bars that have cross-sectional area ε at the optimum. Thereafter nodes that are not connected with a bar anymore are also deleted. This sounds simple, but it can be tricky to find a suitable value for the lower bound ε. If ε is too small, then the equilibrium equations will be solved with large errors due to illconditioning of $\boldsymbol{K}(\boldsymbol{x})$. On the other hand, if ε is too large, then one may end up deleting bars that actually are of great structural importance. In most cases, the truss obtained after this deletion of bars and nodes will have a positive definite stiffness matrix, although a singular stiffness matrix may be obtained in exceptional cases, cf. Fig. 5.5, case c) where the dashed lines now correspond to bars of cross-sectional area ε that are deleted. Note, however, that this structure will have a positive definite stiffness matrix if it were modeled as a one-dimensional truss instead of a two-dimensional one.

5.2.1 Convexity of the Nested Problem

The problem $(\mathbb{P})_{\mathrm{nf}}$ has the nice feature of being convex, cf. Svanberg [33]. This is seen as follows. The derivative of the compliance C, is obtained as

$$\frac{\partial C(\boldsymbol{x})}{\partial x_j} = \boldsymbol{F}^T \frac{\partial \boldsymbol{u}(\boldsymbol{x})}{\partial x_j} = (\boldsymbol{K}(\boldsymbol{x})\boldsymbol{u}(\boldsymbol{x}))^T \frac{\partial \boldsymbol{u}(\boldsymbol{x})}{\partial x_j} = \boldsymbol{u}(\boldsymbol{x})^T \boldsymbol{K}(\boldsymbol{x}) \frac{\partial \boldsymbol{u}(\boldsymbol{x})}{\partial x_j}, \tag{5.19}$$

where we used the fact that $K(x)$ is symmetric. We need to find out $\partial u(x)/\partial x_j$. To that end the equilibrium equations $K(x)u(x) = F$ are differentiated as

$$\frac{\partial K(x)}{\partial x_j} u(x) + K(x) \frac{\partial u(x)}{\partial x_j} = \mathbf{0}, \tag{5.20}$$

which gives

$$\frac{\partial u(x)}{\partial x_j} = -K(x)^{-1} \frac{\partial K(x)}{\partial x_j} u(x) = -K(x)^{-1} K_j^0 u(x), \tag{5.21}$$

where we have used (5.12). From (5.19) and (5.21), we obtain

$$\frac{\partial C(x)}{\partial x_j} = -u(x)^T K_j^0 u(x). \tag{5.22}$$

A second differentiation of the compliance yields

$$\begin{aligned} \frac{\partial^2 C(x)}{\partial x_i \partial x_j} &= -\left(\frac{\partial u(x)}{\partial x_i}\right)^T K_j^0 u(x) - u(x)^T K_j^0 \frac{\partial u(x)}{\partial x_i} \\ &= u(x)^T K_i^0 K(x)^{-1} K_j^0 u(x) + u(x)^T K_j^0 K(x)^{-1} K_i^0 u(x) \\ &= 2u(x)^T K_i^0 K(x)^{-1} K_j^0 u(x), \end{aligned} \tag{5.23}$$

where the last equality holds since

$$\begin{aligned} &u(x)^T K_j^0 K(x)^{-1} K_i^0 u(x) \\ &= (u(x)^T K_j^0 K(x)^{-1} K_i^0 u(x))^T = (K_i^0 u(x))^T (u(x)^T K_j^0 K(x)^{-1})^T \\ &= u(x)^T K_i^0 K(x)^{-1} (u(x)^T K_j^0)^T = u(x)^T K_i^0 K(x)^{-1} K_j^0 u(x). \end{aligned}$$

In order to see that the Hessian $\nabla^2 C(x)$ is positive semidefinite, study

$$\begin{aligned} &y^T \nabla^2 C(x) y \\ &= \sum_{j=1}^n \sum_{i=1}^n \frac{\partial^2 C(x)}{\partial x_i \partial x_j} y_i y_j = 2u(x)^T \left[\sum_{j=1}^n \sum_{i=1}^n K_i^0 y_i K(x)^{-1} K_j^0 y_j \right] u(x). \end{aligned}$$

If we introduce the symmetric matrix $Y = \sum_{i=1}^n K_i^0 y_i$, we get

$$\begin{aligned} y^T \nabla^2 C(x) y &= 2u(x)^T \left[Y K(x)^{-1} Y \right] u(x) \\ &= 2\,(Y u(x))^T K(x)^{-1} (Y u(x)) \geq 0, \end{aligned}$$

where the inequality holds because $K(x)^{-1}$ is positive definite since $K(x)$ is positive definite. Thus, applying Theorem 3.2(i), the compliance is a convex function. Since also the constraint function $\sum_{j=1}^n l_j x_j - V_{\max}$ is convex, $(\mathbb{P})_{\mathrm{nf}}$ is convex.

It should be noted that the simultaneous formulation $(\mathbb{P})_{\mathrm{sf}}$ of the problem is not convex! This nonconvexity is due to the equilibrium equations $\boldsymbol{h}(\boldsymbol{x}, \boldsymbol{u}) = \boldsymbol{K}(\boldsymbol{x})\boldsymbol{u} - \boldsymbol{F} = \boldsymbol{0}$. These equalities may be written as a number of inequalities as $h_i(\boldsymbol{x}, \boldsymbol{u}) \leq 0$ and $-h_i(\boldsymbol{x}, \boldsymbol{u}) \leq 0$, for $i = 1, \dots, n_f$, where n_f is the number of degrees-of-freedom of the truss. In order that the feasible set be convex, both h_i and $-h_i$ need to be convex. This is the case if, and only if, h_i is an affine function of $\boldsymbol{x}$ and $\boldsymbol{u}$, i.e. it may be written as $h_i(\boldsymbol{x}, \boldsymbol{u}) = \boldsymbol{a}^T\boldsymbol{x} + \boldsymbol{b}^T\boldsymbol{u} + c = 0$ for some constant vectors $\boldsymbol{a}$ and $\boldsymbol{b}$, and a constant scalar c. Since the equilibrium equations may not be written on this form, we conclude that $(\mathbb{P})_{\mathrm{sf}}$ is a nonconvex problem.

5.2.2 Fully Stressed Designs

Since problem $(\mathbb{P})_{\mathrm{nf}}$ is convex, the KKT conditions are both necessary and sufficient optimality conditions. In order to write them down, we start by forming the Lagrangian defined in (3.3):

$$\mathcal{L}(\boldsymbol{x}, \lambda) = C(\boldsymbol{x}) + \left(\sum_{i=1}^{n} x_i l_i - V_{\max}\right)\lambda.$$

The derivative of the compliance was obtained in (5.22). This expression may be transformed from the global level to the element level by using (5.11) and (5.9):

$$\begin{aligned}\boldsymbol{u}(\boldsymbol{x})^T\boldsymbol{K}_j^0\boldsymbol{u}(\boldsymbol{x}) &= \boldsymbol{u}(\boldsymbol{x})^T\boldsymbol{C}_j^T\boldsymbol{k}_j^0\boldsymbol{C}_j\boldsymbol{u}(\boldsymbol{x}) = (\boldsymbol{C}_j\boldsymbol{u}(\boldsymbol{x}))^T\boldsymbol{k}_j^0\boldsymbol{C}_j\boldsymbol{u}(\boldsymbol{x}) \\ &= \boldsymbol{u}_j(\boldsymbol{x})^T\boldsymbol{k}_j^0\boldsymbol{u}_j(\boldsymbol{x}),\end{aligned}$$

so that

$$\frac{\partial C(\boldsymbol{x})}{\partial x_j} = -\boldsymbol{u}_j(\boldsymbol{x})^T\boldsymbol{k}_j^0\boldsymbol{u}_j(\boldsymbol{x}). \tag{5.24}$$

Since $\boldsymbol{k}_j^0$ is positive semidefinite, this derivative cannot be positive. This is intuitively clear; if we increase the cross-sectional area of a bar, the stiffness of the truss will increase, i.e. the compliance will decrease. By comparing (5.24) with (5.17) and (5.7), we conclude that the partial derivative of the compliance with respect to the cross-sectional area of bar j equals minus two times the strain energy per unit area of the bar. From (5.16), we rewrite (5.24) as

$$\frac{\partial C(\boldsymbol{x})}{\partial x_j} = -\frac{1}{x_j}E\varepsilon_j^2 x_j l_j = -\frac{\sigma_j^2}{E}l_j.$$

Differentiation of the Lagrangian with respect to x_j yields

$$\frac{\partial \mathcal{L}(\boldsymbol{x}, \lambda)}{\partial x_j} = -\frac{\sigma_j^2}{E}l_j + \lambda l_j.$$

The KKT conditions (3.4)–(3.6) give the following KKT point $(\boldsymbol{x}^*, \lambda^*)$:

$$\sigma_j^2 \le \frac{\lambda^*}{E} \quad \text{if } x_j^* = x_j^{\min}$$

$$\sigma_j^2 = \frac{\lambda^*}{E} \quad \text{if } x_j^{\min} < x_j^* < x_j^{\max}$$

$$\sigma_j^2 \ge \frac{\lambda^*}{E} \quad \text{if } x_j^* = x_j^{\max}.$$

If $x_j^{\max}$, $j = 1, \ldots, n$, are not so small that the maximum volume that the truss can attain is less than $V_{\max}$, then clearly $\lambda^* > 0$, i.e. the volume constraint is active at the optimum, cf. the KKT conditions (3.4)–(3.10). We conclude that all bars where the optimum cross-sectional area is strictly greater than the lower bound and strictly smaller than the upper bound, have the same magnitude of the stress. One therefore speaks of a fully stressed design. Of course, some bars will be in tension and others in compression, i.e. the stress will be positive in some bars and negative in others.

5.2.3 Minimization of the Volume Under a Compliance Constraint

In Sect. 2.4 we minimized the weight of a cantilever under a tip displacement constraint. This problem was then compared with the related problem of minimizing the tip displacement under a weight constraint. It was found that the solution of the latter problem could be obtained by scaling the solution of the former and vice versa. We will now generalize this result for the truss problem at hand. We will denote the problem of minimizing the compliance under a volume constraint (A), and the problem of minimizing the volume under a compliance constraint (B):

$$(A) \quad \begin{cases} \min\limits_{x} \boldsymbol{F}^T \boldsymbol{u}(\boldsymbol{x}) \\ \text{s.t.} \quad \boldsymbol{l}^T \boldsymbol{x} - V_{\max} \le 0 \\ \qquad \boldsymbol{x} \in \mathcal{X} \end{cases} \qquad (B) \quad \begin{cases} \min\limits_{x} \boldsymbol{l}^T \boldsymbol{x} \\ \text{s.t.} \quad \boldsymbol{F}^T \boldsymbol{u}(\boldsymbol{x}) - C_{\max} \le 0 \\ \qquad \boldsymbol{x} \in \mathcal{X}, \end{cases}$$

where $\boldsymbol{l} = [l_1 \; \cdots \; l_n]^T$ is a vector with the lengths of the bars, and $C_{\max} > 0$ is the maximum allowed compliance in problem (B). It is assumed that $\boldsymbol{F} \neq \boldsymbol{0}$.

Under the assumption that the lower and upper bounds in neither (A) nor (B) are active at the solutions, we prove the following: If $\boldsymbol{x}_A^*$ is a solution to (A), then

$$\boldsymbol{x}_B^* = \frac{C_A^*}{C_{\max}} \boldsymbol{x}_A^*, \quad C_A^* = \boldsymbol{F}^T \boldsymbol{u}(\boldsymbol{x}_A^*), \tag{5.25}$$

is a solution to (B). Similarly, if $\boldsymbol{x}_B^*$ is a solution to (B), then

$$\boldsymbol{x}_A^* = \frac{V_{\max}}{V_B^*} \boldsymbol{x}_B^*, \quad V_B^* = \boldsymbol{l}^T \boldsymbol{x}_B^*, \tag{5.26}$$

is a solution to (A). Naturally, when performing these scalings, one may obtain points that are not feasible, and consequently not solutions. As just mentioned, we assume from the beginning that such situations will not be encountered.

Making use of (5.22) for the derivative of the compliance, the KKT conditions for (A) are written

$$-\boldsymbol{u}(\boldsymbol{x}_A)^T \boldsymbol{K}_j^0 \boldsymbol{u}(\boldsymbol{x}_A) + \lambda_A l_j = 0 \tag{5.27}$$

$$\lambda_A(\boldsymbol{l}^T \boldsymbol{x}_A - V_{\max}) = 0 \tag{5.28}$$

$$\boldsymbol{l}^T \boldsymbol{x}_A - V_{\max} \leq 0 \tag{5.29}$$

$$\lambda_A \geq 0, \tag{5.30}$$

and for (B) they become

$$l_j - \lambda_B \boldsymbol{u}(\boldsymbol{x}_B)^T \boldsymbol{K}_j^0 \boldsymbol{u}(\boldsymbol{x}_B) = 0 \tag{5.31}$$

$$\lambda_B(\boldsymbol{F}^T \boldsymbol{u}(\boldsymbol{x}_B) - C_{\max}) = 0 \tag{5.32}$$

$$\boldsymbol{F}^T \boldsymbol{u}(\boldsymbol{x}_B) - C_{\max} \leq 0 \tag{5.33}$$

$$\lambda_B \geq 0. \tag{5.34}$$

Let $\boldsymbol{x}_A^*$ be a solution to (A), and let $\boldsymbol{x}_B^* = (\boldsymbol{F}^T \boldsymbol{u}(\boldsymbol{x}_A^*)/C_{\max})\boldsymbol{x}_A^*$. We will prove that there exists a $\lambda_B^* \geq 0$ such that $(\boldsymbol{x}_B^*, \lambda_B^*)$ is a KKT point of (B), and thus, since the problem is convex, $\boldsymbol{x}_B^*$ is a solution to (B). The following simple lemma will be needed:

Lemma 5.1 *Let the positive definite global stiffness matrix be written as in* (5.12) *for any* $\boldsymbol{x}$, *i.e.* $\boldsymbol{K}(\boldsymbol{x}) = \sum_{j=1}^n x_j \boldsymbol{K}_j^0$, *where* $\boldsymbol{K}_j^0$ *are constant matrices. Let further* $\boldsymbol{x}^* = \alpha \boldsymbol{x}$, $\alpha \neq 0$. *Then* $\boldsymbol{u}(\boldsymbol{x}^*) = \boldsymbol{u}(\boldsymbol{x})/\alpha$ *solves* $\boldsymbol{K}(\boldsymbol{x}^*)\boldsymbol{u}(\boldsymbol{x}^*) = \boldsymbol{F}$ *if, and only if,* $\boldsymbol{u}(\boldsymbol{x})$ *solves* $\boldsymbol{K}(\boldsymbol{x})\boldsymbol{u}(\boldsymbol{x}) = \boldsymbol{F}$.

The lemma follows by rewriting the equilibrium equations for the design $\boldsymbol{x}$, $\boldsymbol{K}(\boldsymbol{x})\boldsymbol{u}(\boldsymbol{x}) = \boldsymbol{F}$, as

$$\sum_{j=1}^n x_j \boldsymbol{K}_j^0 \boldsymbol{u}(\boldsymbol{x}) = \boldsymbol{F} \quad \Longleftrightarrow \quad \sum_{j=1}^n \alpha x_j \boldsymbol{K}_j^0 \left(\frac{\boldsymbol{u}(\boldsymbol{x})}{\alpha}\right) = \boldsymbol{F}$$

$$\Longleftrightarrow \quad \boldsymbol{K}(\boldsymbol{x}^*)\left(\frac{\boldsymbol{u}(\boldsymbol{x})}{\alpha}\right) = \boldsymbol{F}.$$

Thus, $\boldsymbol{u}(\boldsymbol{x}^*) = \boldsymbol{u}(\boldsymbol{x})/\alpha$ is the unique solution to equilibrium equations for the design $\boldsymbol{x}^*$: $\boldsymbol{K}(\boldsymbol{x}^*)\boldsymbol{u}(\boldsymbol{x}^*) = \boldsymbol{F}$.

Using the lemma, we have that

$$\boldsymbol{u}(\boldsymbol{x}_B^*) = \frac{C_{\max}}{\boldsymbol{F}^T \boldsymbol{u}(\boldsymbol{x}_A^*)} \boldsymbol{u}(\boldsymbol{x}_A^*). \tag{5.35}$$

Note that the denominator cannot be zero since the compliance C is always positive: $C = \boldsymbol{F}^T\boldsymbol{u}(\boldsymbol{x}) = \boldsymbol{u}(\boldsymbol{x})^T\boldsymbol{K}(\boldsymbol{x})\boldsymbol{u}(\boldsymbol{x}) > 0$, since $\boldsymbol{K}(\boldsymbol{x})$ is positive definite (otherwise we would not be able to formulate the nested problems we are studying), and $\boldsymbol{u}(\boldsymbol{x}) \neq \boldsymbol{0}$ since we assumed that $\boldsymbol{F} \neq \boldsymbol{0}$. Intuitively, this result is obvious since the compliance is the magnitude of the external force times the displacement in the direction of this force, which certainly must be positive.

From (5.27) we get

$$\lambda_A^* = \frac{\boldsymbol{u}(\boldsymbol{x}_A^*)^T\boldsymbol{K}_j^0\boldsymbol{u}(\boldsymbol{x}_A^*)}{l_j}, \tag{5.36}$$

where $\lambda_A^* > 0$, since if it were zero, then (5.27) implies that the strain energy is zero in all elements, which is impossible. Equations (5.31) and (5.35) give

$$l_j - \lambda_B^*\left(\frac{C_{\max}}{\boldsymbol{F}^T\boldsymbol{u}(\boldsymbol{x}_A^*)}\right)^2\boldsymbol{u}(\boldsymbol{x}_A^*)^T\boldsymbol{K}_j^0\boldsymbol{u}(\boldsymbol{x}_A^*) = 0.$$

Insertion of (5.36) results in

$$\lambda_B^* = \frac{1}{\left(\frac{C_{\max}}{\boldsymbol{F}^T\boldsymbol{u}(\boldsymbol{x}_A^*)}\right)^2\lambda_A^*} > 0,$$

so (5.34) is satisfied. From (5.35) we have

$$\boldsymbol{F}^T\boldsymbol{u}(\boldsymbol{x}_B^*) - C_{\max} = \boldsymbol{F}^T\frac{C_{\max}}{\boldsymbol{F}^T\boldsymbol{u}(\boldsymbol{x}_A^*)}\boldsymbol{u}(\boldsymbol{x}_A^*) - C_{\max} = 0,$$

and thus, (5.32) and (5.33) are satisfied. Since all KKT conditions of (B) are satisfied, we know that $\boldsymbol{x}_B^*$ as defined above is a solution to (B).

Conversely, let $\boldsymbol{x}_B^*$ be a solution to (B), and let $\boldsymbol{x}_A^* = (V_{\max}/\boldsymbol{l}^T\boldsymbol{x}_B^*)\boldsymbol{x}_B^*$. Lemma 5.1 then implies that

$$\boldsymbol{u}(\boldsymbol{x}_A^*) = \frac{\boldsymbol{l}^T\boldsymbol{x}_B^*}{V_{\max}}\boldsymbol{u}(\boldsymbol{x}_B^*).$$

Combining this with (5.27) and (5.31) gives

$$\lambda_A^* = \frac{\left(\frac{\boldsymbol{l}^T\boldsymbol{x}_B^*}{V_{\max}}\right)^2}{\lambda_B^*} > 0,$$

so (5.30) is valid. Finally, (5.26) yields

$$\boldsymbol{l}^T\boldsymbol{x}_A^* - V_{\max} = \boldsymbol{l}^T\left(\frac{V_{\max}}{\boldsymbol{l}^T\boldsymbol{x}_B^*}\right)\boldsymbol{x}_B^* - V_{\max} = 0.$$

Thus, also (5.28) and (5.29) hold true, proving that $\boldsymbol{x}_A^*$ is indeed a solution to (A).

5.3 Numerical Solution of the Nested Problem Using MMA

We now turn to the problem of solving $(\mathbb{P})_{\mathrm{nf}}$ numerically by generating a sequence of explicit, convex approximations. As proven above, $(\mathbb{P})_{\mathrm{nf}}$ is in fact convex, but that in no way lessens the desire to use approximations that are both explicit and convex. We will use an MMA approximation of the compliance, $\hat{g}_0(\boldsymbol{x}) = g_0(\boldsymbol{x}, \boldsymbol{u}(\boldsymbol{x})) = \boldsymbol{F}^T \boldsymbol{u}(\boldsymbol{x})$. To that end we need the derivative of the compliance. This has been calculated in (5.24) as

$$\frac{\partial \hat{g}_0(\boldsymbol{x})}{\partial x_j} = -\boldsymbol{u}_j(\boldsymbol{x})^T \boldsymbol{k}_j^0 \boldsymbol{u}_j(\boldsymbol{x}).$$

Since $\boldsymbol{k}_j^0$ is positive semidefinite, we have $\partial \hat{g}_0(\boldsymbol{x})/\partial x_j \leq 0$. With this information, the MMA approximation of $\hat{g}_0(\boldsymbol{x})$ at the design $\boldsymbol{x}^k$ may be formulated. From (4.8), (4.10) and (4.11) we get

$$\hat{g}_0^{M,k}(\boldsymbol{x}) = r_0^k + \sum_{j=1}^{n} \frac{q_{0j}^k}{x_j - L_j^k},$$

where

$$q_{0j}^k = (x_j^k - L_j^k)^2 \boldsymbol{u}_j(\boldsymbol{x}^k)^T \boldsymbol{k}_j^0 \boldsymbol{u}_j(\boldsymbol{x}^k) \tag{5.37}$$

$$r_0^k = g_0(\boldsymbol{x}^k) - \sum_{j=1}^{n} (x_j^k - L_j^k) \boldsymbol{u}_j(\boldsymbol{x}^k)^T \boldsymbol{k}_j^0 \boldsymbol{u}_j(\boldsymbol{x}^k). \tag{5.38}$$

Note that $\hat{g}_0$ is linearized in the variables $1/(x_j - L_j^k)$ only; there are no terms $1/(U_j^k - x_j)$ present. This is because $p_{0j}^k = 0$ in (4.9) since $\partial \hat{g}_0(\boldsymbol{x}^k)/\partial x_j \leq 0$.

Since the volume is a linear function of the design variable, we choose not to approximate the volume constraint. Thus, the approximation of $(\mathbb{P})_{\mathrm{nf}}$ for iteration k is

$$(\mathbb{P})_{\mathrm{nf}}^{M,k} \quad \begin{cases} \min\limits_{\boldsymbol{x}} \hat{g}_0^{M,k}(\boldsymbol{x}) \\ \text{s.t.} \quad \hat{g}_1(\boldsymbol{x}) = \sum\limits_{j=1}^{n} l_j x_j - V_{\max} \leq 0 \\ \qquad \alpha_j^k \leq x_j \leq x_j^{\max}, \quad j = 1, \ldots, n, \end{cases}$$

where α_j^k are move limits defined as $\alpha_j^k = \max(x_j^{\min},\ L_j^k + \mu(x_j^k - L_j^k))$, where $0 < \mu < 1$. The purpose of introducing α_j^k is to prevent division by zero in $\hat{g}_0^{M,k}(\boldsymbol{x})$ since they ensure that $L_j^k < \alpha_j^k \leq x_j^k$. If all $q_{0j}^k > 0$, i.e. the strain energy of all bars is nonzero, then $\hat{g}_0^{M,k}$ is strictly convex, and thus there is at most one solution to $(\mathbb{P})_{\mathrm{nf}}^{M,k}$; it could happen that there are no feasible points, and thus, no solution.

The subproblem $(\mathbb{P})_{\text{nf}}^{M,k}$ is easily solved by using Lagrangian duality. The Lagrangian is

$$\mathcal{L}^k(\boldsymbol{x},\lambda) = \hat{g}_0^{M,k}(\boldsymbol{x}) + \lambda\left(\sum_{j=1}^{n} l_j x_j - V_{\max}\right)$$

$$= r_0^k + \sum_{j=1}^{n}\left(\frac{q_{0j}^k}{x_j - L_j^k} + \lambda l_j x_j\right) - \lambda V_{\max}.$$

The dual objective function is

$$\varphi^k(\lambda) = \min_{\boldsymbol{x}} \mathcal{L}^k(\boldsymbol{x},\lambda) = r_0^k - \lambda V_{\max} + \sum_{j=1}^{n} \min_{\substack{\alpha_j^k \le x_j \\ \le x_j^{\max}}} \underbrace{\left(\frac{q_{0j}^k}{x_j - L_j^k} + \lambda l_j x_j\right)}_{\mathcal{L}_j^k(x_j,\lambda)},$$

where we have taken advantage of the separability of $\mathcal{L}^k$, cf. Sect. 3.4.1. The function $x_j \mapsto \mathcal{L}_j(x_j,\boldsymbol{\lambda})$ is strictly convex (unless $q_{0j}^k = 0$). In order to minimize $\mathcal{L}_j^k$ with respect to x_j, we first make the guess that the minimum is obtained for the x_j-value, denoted x_j^t (t for "trial"), for which the partial derivative of $\mathcal{L}_j^k$ with respect to x_j is zero:

$$\frac{\partial \mathcal{L}_j^k(x_j,\lambda)}{\partial x_j} = -\frac{q_{0j}^k}{(x_j - L_j^k)^2} + \lambda l_j = 0,$$

which gives

$$x_j = x_j^t = L_j^k + \sqrt{\frac{q_{0j}^k}{\lambda l_j}}.$$

If $x_j^t < \alpha_j^k$, we conclude that $x_j^* = \alpha_j^k$ minimizes $\mathcal{L}_j^k$. Similarly, if $x_j^t > x_j^{\max}$, then $x_j^* = x_j^{\max}$. To summarize, we have

$$x_j^*(\lambda) = \begin{cases} \alpha_j^k & \text{if } x_j^t < \alpha_j^k \\ x_j^{\max} & \text{if } x_j^t > x_j^{\max} \\ x_j^t & \text{otherwise.} \end{cases} \tag{5.39}$$

Finally, the dual problem of $(\mathbb{P})_{\text{nf}}^{M,k}$ may be written

$$(D)^k \quad \begin{cases} \max\limits_{\lambda} \varphi^k(\lambda) \\ \text{s.t.} \quad \lambda \ge 0. \end{cases}$$

Since φ^k is a concave function in only one variable, $(D)^k$ is very easy to solve using some iterative nonlinear programming method, such as the Golden Section method, the method of steepest descent, or Newton's method.

We end this section by summarizing the procedure to solve $(\mathbb{P})_{\text{nf}}$ using a sequence of MMA approximated subproblems.

1. As input the following is needed: The location of the nodes, what nodes that are (partially) fixed, the applied force $\boldsymbol{F}$, Young's modulus E, lower and upper bounds, $x_j^{\min}$ and $x_j^{\max}$, on the cross-sectional areas, and the maximum allowed volume of the truss, $V_{\max}$. In addition, an initial design $\boldsymbol{x}^0$, initial values of the lower asymptotes, L_j^0, as well as an initial value $\lambda > 0$ of the dual variable. Put the iteration index $k = 0$.
2. Perform a finite element analysis in order to obtain the displacement vector for design $\boldsymbol{x}^k$:
$$\boldsymbol{u}(\boldsymbol{x}^k) = \boldsymbol{K}(\boldsymbol{x}^k)^{-1}\boldsymbol{F}.$$
3. Calculate the objective function $\hat{g}_0(\boldsymbol{x}^k) = \boldsymbol{F}^T\boldsymbol{u}(\boldsymbol{x}^k)$. Perform a sensitivity analysis by calculating
$$\frac{\partial \hat{g}_0(\boldsymbol{x}^k)}{\partial x_j} = -\boldsymbol{u}_j(\boldsymbol{x}^k)^T\boldsymbol{k}_j^0\boldsymbol{u}_j(\boldsymbol{x}^k),$$
for all bars $j = 1, \ldots, n$.
4. Formulate an MMA approximation at $\boldsymbol{x}^k$ by calculating q_{0j}^k and r_0^k in (5.37) and (5.38) for all bars $j = 1, \ldots, n$.
5. Solve the dual problem $(D)^k$ iteratively by some nonlinear optimization algorithm. In each iteration of this algorithm, the dual objective function is calculated as
$$\varphi^k(\lambda) = r_0^k - \lambda V_{\max} + \sum_{j=1}^{n}\left(\frac{q_{0j}^k}{x_j^*(\lambda) - L_j^k} + \lambda l_j x_j^*(\lambda)\right),$$
for the current value of λ. Here, $x_j^*(\lambda)$, $j = 1, \ldots, n$, is obtained from (5.39). If a gradient method is used to solve $(D)^k$, then calculate the gradient of φ^k from (3.13) as
$$\frac{\partial \varphi^k(\lambda)}{\partial \lambda} = g_1(\boldsymbol{x}^*(\lambda)) = \sum_{j=1}^{n} l_j x_j^*(\lambda) - V_{\max}.$$
(If a Newton method is used, $\partial^2\varphi^k(\lambda)/\partial\lambda^2$ needs to be calculated.)
   When the solution of $(D)^k$ has been found within some tolerance, put the current, i.e. the optimal, value of λ to λ^*. Calculate the corresponding optimum design variables $x_j^*(\lambda^*)$, $j = 1, \ldots, n$, from (5.39). Put $\boldsymbol{x}^{k+1} = \boldsymbol{x}^*$.
6. If $\boldsymbol{x}^{k+1}$ is considered a sufficiently good solution to $(\mathbb{P})_{\text{nf}}$ (e.g. because the objective function and the design variables have not changed much from the previous iteration), then stop. If not, put $k = k + 1$ and update the lower asymptotes L_j^k and the move limits α_j^k according to some heuristic rule, and return to step 2 for a new iteration.

The most time-consuming task in this algorithm is the FE-analysis in step 2.

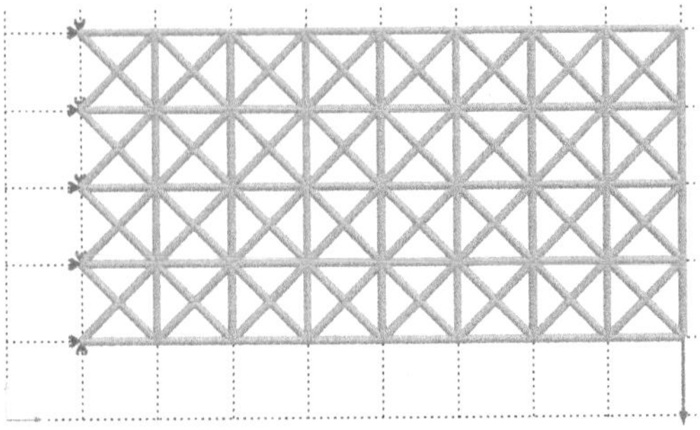

Fig. 5.6 (Color online) The initial configuration of the truss

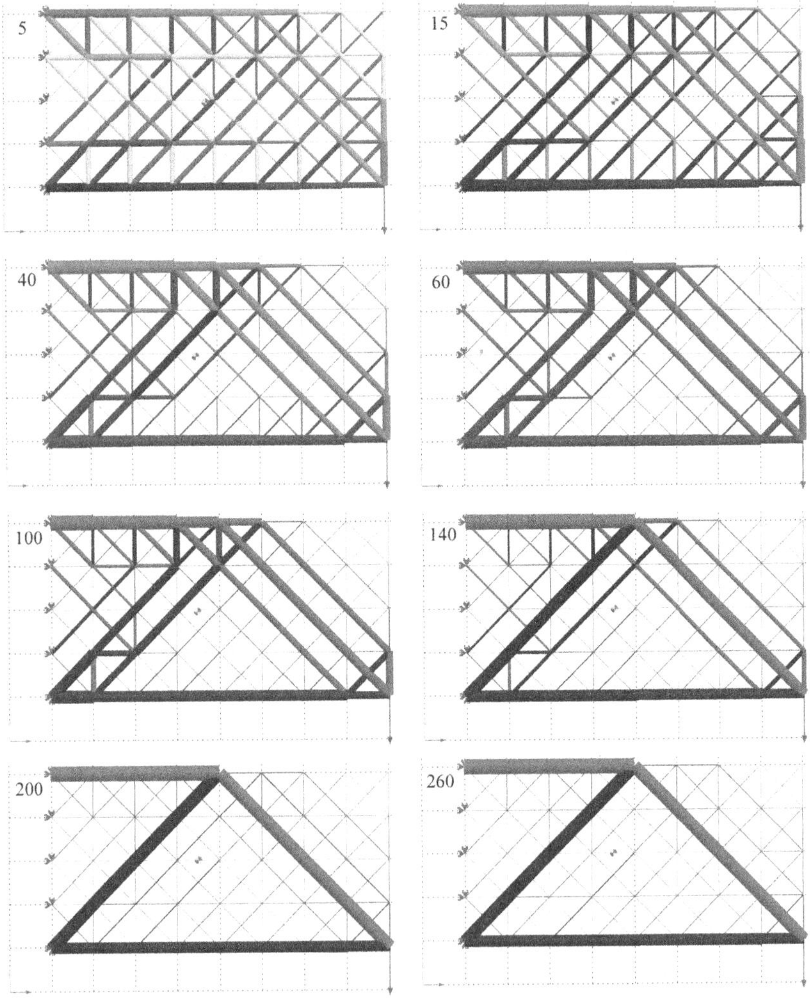

Fig. 5.7 (Color online) The truss after 5, 15, 40, 60, 100, 140, 200 and 260 iterations. The thickness of each bar in the plot is directly proportional to its cross-sectional area

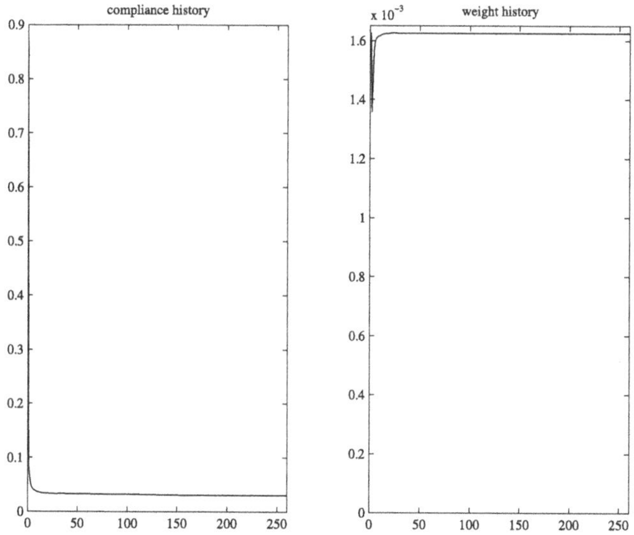

Fig. 5.8 The history of the compliance and the weight during the iterations

Example 5.2 The cross-sectional areas that minimize the compliance of the 136-bar truss in Fig. 5.6 should be determined. The truss is subjected to the single applied force shown in the figure, and all bars have the same density. Initially, all bars have the cross-sectional area 10^{-4} units. The volume, or, equivalently, the weight, of the truss is not allowed to increase from the initial value. The minimum and maximum allowed values of the cross-sectional area are 10^{-6} and 10^{-2} units, respectively. Figure 5.7 shows the truss at various iterations of the algorithm described above as implemented in the finite element program TRINITAS. Effectively, the truss converges to a four-bar truss. Note, from Fig. 5.8, that the compliance changes almost nothing from around iteration number 20, although there are significant changes of the design variables up to around iteration number 200.

Chapter 6
Sensitivity Analysis

When solving nested structural optimization problems by generating a sequence of explicit first order approximations, such as MMA, one needs to differentiate the objective function and all constraint functions with respect to the design variables. The procedure to obtain these derivatives, or sensitivities, is called sensitivity analysis. In the previous chapter we determined the sensitivity of the compliance of a truss with respect to the cross-sectional area of the bars. In this chapter we will go further and describe how to perform a sensitivity analysis for arbitrary functions and design variables. There are two main groups of methods: numerical methods, which are all approximate, and analytical methods, which are exact. One may also consider hybrids of methods from these two groups: so-called semianalytical methods.

6.1 Numerical Methods

We recall that the nested structural optimization problem may be written as

$$(\mathbb{SO})_{\text{nf}} \quad \begin{cases} \min_{\boldsymbol{x}} \hat{g}_0(\boldsymbol{x}) = g_0(\boldsymbol{x}, \boldsymbol{u}(\boldsymbol{x})) \\ \text{s.t.} \quad \hat{g}_i(\boldsymbol{x}) = g_i(\boldsymbol{x}, \boldsymbol{u}(\boldsymbol{x})) \leq 0, \quad i = 1, \ldots, l \\ \qquad \boldsymbol{x} \in \mathcal{X} = \{\boldsymbol{x} \in \mathbb{R}^n : x_j^{\min} \leq x_j \leq x_j^{\max}, \ j = 1, \ldots, n\}, \end{cases}$$

where $\boldsymbol{x} \mapsto \boldsymbol{u}(\boldsymbol{x})$ is an implicit function defined through the equilibrium equations $\boldsymbol{K}(\boldsymbol{x})\boldsymbol{u}(\boldsymbol{x}) = \boldsymbol{F}(\boldsymbol{x})$. In numerical sensitivity analysis methods, $\partial \hat{g}_i / \partial x_j$, $i = 0, \ldots, l$, are approximated by finite differences, e.g. forward or central differences.

The forward difference approximation of $\partial \hat{g}_i / \partial x_j$ at a design $\boldsymbol{x}^k$ is

$$\frac{\partial \hat{g}_i(\boldsymbol{x}^k)}{\partial x_j} \approx D_f = \frac{\hat{g}_i(\boldsymbol{x}^k + h\boldsymbol{e}_j) - \hat{g}_i(\boldsymbol{x}^k)}{h}, \tag{6.1}$$

where $\boldsymbol{e}_j = [0, \ldots, 0, 1, 0, \ldots, 0]^T$ and the 1 is in row j.

In order to illustrate the method, let us see how to calculate, at a design $\boldsymbol{x}^k$, the sensitivity of the compliance $\hat{g}_0(\boldsymbol{x}) = g_0(\boldsymbol{x}, \boldsymbol{u}(\boldsymbol{x})) = \boldsymbol{F}(\boldsymbol{x})^T \boldsymbol{u}(\boldsymbol{x})$ with respect to changes in the cross-sectional area $x_j = A_j$ of bar j in a truss. First the compliance $\hat{g}_0(\boldsymbol{x}^k)$ is calculated. Then, a small number $h > 0$ is added to the cross-sectional area of bar j. We need to find the compliance for this new design $\boldsymbol{x}^k + h\boldsymbol{e}_j$. In order to do so we first have to calculate the displacement for this design, i.e. perform a finite element analysis to solve the equilibrium equations. Once $\boldsymbol{u}(\boldsymbol{x}^k + h\boldsymbol{e}_j)$ is known, the compliance $\hat{g}_0(\boldsymbol{x}^k + h\boldsymbol{e}_j)$ may be calculated. Insertion into (6.1) then gives the desired sensitivity. A major problem with this method is that it can be difficult to find a suitable h. Clearly, if h is too large, then D_f will be a bad approximation of $\partial \hat{g}_i(\boldsymbol{x}^k)/\partial x_j$. It may be shown that the truncation error $\partial \hat{g}_i(\boldsymbol{x}^k)/\partial x_j - D_f = O(h)$

as $h \to 0$. However, as $h \to 0$, the numerical error due to cancellation increases dramatically, so one should not choose h too small either.

The central difference approximation of $\partial \hat{g}_i / \partial x_j$ at $\boldsymbol{x}^k$ is defined as

$$\frac{\partial \hat{g}_i(\boldsymbol{x}^k)}{\partial x_j} \approx \frac{\hat{g}_i(\boldsymbol{x}^k + h\boldsymbol{e}_j) - \hat{g}_i(\boldsymbol{x}^k - h\boldsymbol{e}_j)}{2h}. \tag{6.2}$$

This approximation is more accurate; the truncation error is $O(h^2)$ as $h \to 0$. On the other hand it is more expensive, since one has to perform n additional finite element analyses for the designs $\boldsymbol{x}^k - h\boldsymbol{e}_j$.

Although numerical sensitivity analysis methods may be very inaccurate and expensive, they at least have one advantage: they are very easy to implement.

6.2 Analytical Methods

In order to obtain analytical expressions for $\partial \hat{g}_i(\boldsymbol{x}^k)/\partial x_j$, the chain rule is first applied:

$$\frac{\partial \hat{g}_i(\boldsymbol{x}^k)}{\partial x_j} = \frac{\partial g_i(\boldsymbol{x}^k, \boldsymbol{u}(\boldsymbol{x}^k))}{\partial x_j} + \frac{\partial g_i(\boldsymbol{x}^k, \boldsymbol{u}(\boldsymbol{x}^k))}{\partial \boldsymbol{u}} \frac{\partial \boldsymbol{u}(\boldsymbol{x}^k))}{\partial x_j}, \tag{6.3}$$

where $\partial g_i / \partial \boldsymbol{u}$ is a row matrix, and $\partial \boldsymbol{u} / \partial x_j$ is a column matrix. Two different analytical methods will be studied: the so-called direct and adjoint methods.

6.2.1 Direct Analytical Method

In the direct analytical method, $\partial \boldsymbol{u}(\boldsymbol{x}^k)/\partial x_j$ is obtained by differentiation of the equilibrium equations $\boldsymbol{K}(\boldsymbol{x})\boldsymbol{u}(\boldsymbol{x}) = \boldsymbol{F}(\boldsymbol{x})$. The result is then inserted into (6.3). We get

$$\frac{\partial \boldsymbol{K}(\boldsymbol{x}^k)}{\partial x_j}\boldsymbol{u}(\boldsymbol{x}^k) + \boldsymbol{K}(\boldsymbol{x}^k)\frac{\partial \boldsymbol{u}(\boldsymbol{x}^k)}{\partial x_j} = \frac{\partial \boldsymbol{F}(\boldsymbol{x}^k)}{\partial x_j},$$

which is rewritten as

$$\boldsymbol{K}(\boldsymbol{x}^k)\frac{\partial \boldsymbol{u}(\boldsymbol{x}^k)}{\partial x_j} = \frac{\partial \boldsymbol{F}(\boldsymbol{x}^k)}{\partial x_j} - \frac{\partial \boldsymbol{K}(\boldsymbol{x}^k)}{\partial x_j}\boldsymbol{u}(\boldsymbol{x}^k). \tag{6.4}$$

In order to find $\partial \boldsymbol{u}(\boldsymbol{x}^k)/\partial x_j$, we first need expressions for $\partial \boldsymbol{F}(\boldsymbol{x}^k)/\partial x_j$ and $(\partial \boldsymbol{K}(\boldsymbol{x}^k)/\partial x_j)\boldsymbol{u}(\boldsymbol{x}^k)$. How this may be accomplished will be described shortly. Note that (6.4) has the same structure as the equilibrium equations $\boldsymbol{K}(\boldsymbol{x}^k)\boldsymbol{u}(\boldsymbol{x}^k) = \boldsymbol{F}(\boldsymbol{x}^k)$ that have already been solved. For this reason, the right-hand side in (6.4) is often denoted *pseudo-load*. If the equilibrium equations have been solved by a direct solver rather than with an iterative one, so that $\boldsymbol{K}(\boldsymbol{x}^k)$ has been factorized (for instance by performing a Cholesky decomposition, i.e. a nonsingular lower left triangular matrix $\boldsymbol{L}$ has been found such that $\boldsymbol{K}(\boldsymbol{x}^k) = \boldsymbol{L}\boldsymbol{L}^T$), then only forward and backward substitutions are needed to solve (6.4) for $\partial \boldsymbol{u}(\boldsymbol{x}^k)/\partial x_j$. Thus, it is computationally much cheaper to solve (6.4) for a certain design variable x_j than the equilibrium equations. On the other hand, (6.4) needs to be solved n times, so the amount of time needed to calculate the sensitivity for all design variables may still be considerable if there is a large number of design variables.

Example 6.1 The sensitivity of the compliance, $\hat{g}_0(\boldsymbol{x}) = g_0(\boldsymbol{x}, \boldsymbol{u}(\boldsymbol{x})) = \boldsymbol{F}(\boldsymbol{x})^T \boldsymbol{u}(\boldsymbol{x})$, should be calculated using the direct method. Differentiation of g_0 gives

$$\frac{\partial g_0(\boldsymbol{x}^k, \boldsymbol{u}(\boldsymbol{x}^k))}{\partial x_j} = \frac{\partial \boldsymbol{F}(\boldsymbol{x}^k)^T}{\partial x_j} \boldsymbol{u}(\boldsymbol{x}^k), \qquad \frac{\partial g_0(\boldsymbol{x}^k, \boldsymbol{u}(\boldsymbol{x}^k))}{\partial \boldsymbol{u}} = \boldsymbol{F}(\boldsymbol{x}^k)^T.$$

From (6.4),

$$\frac{\partial \boldsymbol{u}(\boldsymbol{x}^k)}{\partial x_j} = \boldsymbol{K}(\boldsymbol{x}^k)^{-1} \left(\frac{\partial \boldsymbol{F}(\boldsymbol{x}^k)}{\partial x_j} - \frac{\partial \boldsymbol{K}(\boldsymbol{x}^k)}{\partial x_j} \boldsymbol{u}(\boldsymbol{x}^k) \right).$$

Insertion of this into (6.3) yields

$$\begin{aligned} \frac{\partial \hat{g}_0(\boldsymbol{x}^k)}{\partial x_j} &= \frac{\partial \boldsymbol{F}(\boldsymbol{x}^k)^T}{\partial x_j} \boldsymbol{u}(\boldsymbol{x}^k) + \boldsymbol{F}(\boldsymbol{x}^k)^T \boldsymbol{K}(\boldsymbol{x}^k)^{-1} \left(\frac{\partial \boldsymbol{F}(\boldsymbol{x}^k)}{\partial x_j} - \frac{\partial \boldsymbol{K}(\boldsymbol{x}^k)}{\partial x_j} \boldsymbol{u}(\boldsymbol{x}^k) \right) \\ &= 2\boldsymbol{u}(\boldsymbol{x}^k)^T \frac{\partial \boldsymbol{F}(\boldsymbol{x}^k)}{\partial x_j} - \boldsymbol{u}(\boldsymbol{x}^k)^T \frac{\partial \boldsymbol{K}(\boldsymbol{x}^k)}{\partial x_j} \boldsymbol{u}(\boldsymbol{x}^k). \end{aligned} \tag{6.5}$$

6.2.2 Adjoint Analytical Method

If (6.4) is substituted into (6.3), one obtains

$$\frac{\partial \hat{g}_i(\boldsymbol{x}^k)}{\partial x_j} = \frac{\partial g_i}{\partial x_j} + \frac{\partial g_i}{\partial \boldsymbol{u}} \boldsymbol{K}(\boldsymbol{x}^k)^{-1} \left(\frac{\partial \boldsymbol{F}(\boldsymbol{x}^k)}{\partial x_j} - \frac{\partial \boldsymbol{K}(\boldsymbol{x}^k)}{\partial x_j} \boldsymbol{u}(\boldsymbol{x}^k) \right), \tag{6.6}$$

where $g_i = g_i(\boldsymbol{x}^k, \boldsymbol{u}(\boldsymbol{x}^k))$. In this expression, we define

$$\boldsymbol{\lambda}_i = \left(\frac{\partial g_i}{\partial \boldsymbol{u}} \boldsymbol{K}(\boldsymbol{x}^k)^{-1} \right)^T = \boldsymbol{K}(\boldsymbol{x}^k)^{-1} \left(\frac{\partial g_i}{\partial \boldsymbol{u}} \right)^T.$$

In the adjoint method, one starts by solving

$$\boldsymbol{K}(\boldsymbol{x}^k)\boldsymbol{\lambda}_i = \left(\frac{\partial g_i}{\partial \boldsymbol{u}} \right)^T \tag{6.7}$$

for $\boldsymbol{\lambda}_i$. This is then inserted into (6.6) to give the desired sensitivity as

$$\frac{\partial \hat{g}_i(\boldsymbol{x}^k)}{\partial x_j} = \frac{\partial g_i}{\partial x_j} + \boldsymbol{\lambda}_i^T \left(\frac{\partial \boldsymbol{F}(\boldsymbol{x}^k)}{\partial x_j} - \frac{\partial \boldsymbol{K}(\boldsymbol{x}^k)}{\partial x_j} \boldsymbol{u}(\boldsymbol{x}^k) \right). \tag{6.8}$$

At this point we may compare the direct and the adjoint method to calculate $\partial \hat{g}_i(\boldsymbol{x}^k)/\partial x_j$ for $j = 1, \ldots, n$ and $i = 0, \ldots, l$. In the direct method, one needs to solve (6.4) once for each design variable x_j, i.e. n times. The result is then inserted into (6.3) $l+1$ times for each j. In the adjoint method, (6.7) is solved for the objective function and each constraint function, i.e. $l+1$ times. The result is then inserted into (6.8) n times for each i $(= 0, \ldots, l)$. Thus, we conclude that the adjoint method is to be preferred when there are fewer constraints than design variables, otherwise the direct method will be more efficient.

Example 6.2 Let us continue Example 6.1 by this time using the adjoint method to calculate the sensitivity of the compliance. Equation (6.7) reads

$$\boldsymbol{K}(\boldsymbol{x}^k)\boldsymbol{\lambda} = \left(\frac{g_0(\boldsymbol{x}^k, \boldsymbol{u}(\boldsymbol{x}^k))}{\partial \boldsymbol{u}} \right)^T = \boldsymbol{F}(\boldsymbol{x}^k).$$

This system of equations is identical to the equilibrium equations. Thus, since $\boldsymbol{K}(\boldsymbol{x}^k)$ is nonsingular, we conclude that $\boldsymbol{\lambda} = \boldsymbol{u}(\boldsymbol{x}^k)$! Obviously, there is no need to solve (6.7) in this case since we already have $\boldsymbol{u}(\boldsymbol{x}^k)$ available. Insertion into (6.8) gives

$$\begin{aligned} \frac{\partial \hat{g}_0(\boldsymbol{x}^k)}{\partial x_j} &= \frac{\partial \boldsymbol{F}(\boldsymbol{x}^k)^T}{\partial x_j} \boldsymbol{u}(\boldsymbol{x}^k) + \boldsymbol{u}(\boldsymbol{x}^k)^T \left(\frac{\partial \boldsymbol{F}(\boldsymbol{x}^k)}{\partial x_j} - \frac{\partial \boldsymbol{K}(\boldsymbol{x}^k)}{\partial x_j} \boldsymbol{u}(\boldsymbol{x}^k) \right) \\ &= 2\boldsymbol{u}(\boldsymbol{x}^k)^T \frac{\partial \boldsymbol{F}(\boldsymbol{x}^k)}{\partial x_j} - \boldsymbol{u}(\boldsymbol{x}^k)^T \frac{\partial \boldsymbol{K}(\boldsymbol{x}^k)}{\partial x_j} \boldsymbol{u}(\boldsymbol{x}^k). \end{aligned}$$

Thus, as expected, the results of the direct and adjoint methods coincide.

We end this section with some brief words on semianalytical methods. If finite difference approximations are used for $\partial \boldsymbol{u}(\boldsymbol{x}^k)/\partial x_j$ in (6.3), or for $\partial \boldsymbol{F}(\boldsymbol{x}^k)/\partial x_j$ or $\partial \boldsymbol{K}(\boldsymbol{x}^k)/\partial x_j$ in the direct or the adjoint method, then a semianalytical method is obtained. These methods are better than the fully numerical methods since parts of the sensitivity analysis are performed without approximations. In the next section we will see that it is actually fairly easy, and, more importantly, numerically quite inexpensive, to obtain analytical sensitivities of $\boldsymbol{F}$ and $\boldsymbol{K}$. Thus, numerical and semianalytical sensitivity analysis methods are of limited value.

6.3 Analytical Calculation of Pseudo-loads

The pseudo-loads appearing in (6.4) and (6.8) may be calculated by adding the contributions from each element in the structure. Equations (5.14), (5.11) and (5.9) give

$$\begin{aligned}\frac{\partial \boldsymbol{F}(\boldsymbol{x})}{\partial x_j} - \frac{\partial \boldsymbol{K}(\boldsymbol{x})}{\partial x_j}\boldsymbol{u}(\boldsymbol{x}) &= \sum_{e=1}^{n}\left[\boldsymbol{C}_e^T \frac{\partial \boldsymbol{f}_e^a(\boldsymbol{x})}{\partial x_j} - \frac{\partial \boldsymbol{K}_e(\boldsymbol{x})}{\partial x_j}\boldsymbol{u}(\boldsymbol{x})\right] \\ &= \sum_{e=1}^{n}\left[\boldsymbol{C}_e^T \frac{\partial \boldsymbol{f}_e^a(\boldsymbol{x})}{\partial x_j} - \boldsymbol{C}_e^T \frac{\partial \boldsymbol{k}_e(\boldsymbol{x})}{\partial x_j}\boldsymbol{C}_e\boldsymbol{u}(\boldsymbol{x})\right] \\ &= \sum_{e=1}^{n}\boldsymbol{C}_e^T\left[\frac{\partial \boldsymbol{f}_e^a(\boldsymbol{x})}{\partial x_j} - \frac{\partial \boldsymbol{k}_e(\boldsymbol{x})}{\partial x_j}\boldsymbol{u}_e(\boldsymbol{x})\right],\end{aligned}$$

where it has been assumed that $\boldsymbol{C}_e$ is independent of $\boldsymbol{x}$, i.e. design changes do not affect which elements are interconnected or which nodes are suppressed. Thus, we may obtain the pseudo-loads by assembling the sensitivity of the applied load vector and the sensitivity of the element stiffness matrix times the element displacements, in exactly the same way as the applied load vector is assembled in an ordinary finite element analysis to solve the equilibrium equations. Using the assembly operator, cf. (5.15), we may therefore write

$$\frac{\partial \boldsymbol{F}(\boldsymbol{x})}{\partial x_j} - \frac{\partial \boldsymbol{K}(\boldsymbol{x})}{\partial x_j}\boldsymbol{u}(\boldsymbol{x}) = \mathop{\mathbf{A}}_{e=1}^{n}\left[\frac{\partial \boldsymbol{f}_e^a(\boldsymbol{x})}{\partial x_j} - \frac{\partial \boldsymbol{k}_e(\boldsymbol{x})}{\partial x_j}\boldsymbol{u}_e(\boldsymbol{x})\right]. \tag{6.9}$$

It should be pointed out that although we described the assembly of finite elements for a truss only, all equations referred to in the derivation of (6.9) remain valid for any type of displacement-based finite elements.

In the following sections the calculation of analytical expressions for the element-wise contributions to the pseudo-loads will be exemplified by studying bars and sheets.

6.3.1 Bars

The element stiffness matrix $\boldsymbol{k}_e$ for a general bar e in a plane truss, i.e. a truss where all bars lie in the same plane, was derived in Sect. 5.1. It was found that

$$\boldsymbol{k}_e = \boldsymbol{B}_e^T D_e \boldsymbol{B}_e, \tag{6.10}$$

where

$$D_e = \frac{E_e A_e}{l_e}, \tag{6.11}$$

is a scalar relating the elongation δ_e (generalized strain) to the bar force s_e (generalized stress) as $s_e = D_e\delta_e$. E_e is Young's modulus, A_e the cross-sectional area and l_e the length of bar e. $\boldsymbol{B}_e$ is the (generalized) strain displacement matrix

$$\boldsymbol{B}_e = \begin{bmatrix}-c & -s & c & s\end{bmatrix}, \tag{6.12}$$

where $s = \sin\theta_e$, $c = \cos\theta_e$, and the angle θ_e determines the orientation of the bar, cf. Fig. 5.2. It holds that $\delta_e = \boldsymbol{B}_e\boldsymbol{u}_e$ where $\boldsymbol{u}_e$ are the displacements of the end points of the bar. Explicitly, $\boldsymbol{k}_e$ takes the form

$$\boldsymbol{k}_e = \frac{E_e A_e}{l_e}\begin{bmatrix} c^2 & sc & -c^2 & -sc \\ sc & s^2 & -sc & -s^2 \\ -sc & -c^2 & c^2 & sc \\ -sc & -s^2 & sc & s^2 \end{bmatrix}. \tag{6.13}$$

The sensitivity of the stiffness matrix will depend on the nature of the design variables.

6.3.1.1 Sizing and Topology Optimization

For sizing and topology optimization, the cross-sectional areas are the design variables: $x_e = A_e$. The only difference between sizing and topology optimization in this case, is that for topology optimization, the areas are allowed to become zero so that bars may disappear. In this type of problem, l_e and θ_e are constant. Thus, $\boldsymbol{B}_e$ is constant, whereas $D_e = D_e(x_e) = E_e x_e / l_e$. From (6.10) we obtain

$$\frac{\partial \boldsymbol{k}_e}{\partial x_j} = \begin{cases} \dfrac{E_e}{l_e}\boldsymbol{B}_e^T\boldsymbol{B}_e, & \text{if } j = e \\ \boldsymbol{0}, & \text{otherwise.} \end{cases}$$

6.3.1.2 Shape Optimization

By shape optimization of bars we understand that the locations of the end points of the bars are taken as the design variables $\boldsymbol{x}$, see Fig. 6.1. This means that $l_e = l_e(\boldsymbol{x})$, $\theta_e = \theta_e(\boldsymbol{x})$, whereas A_e is constant. Differentiation of (6.10) then results in

$$\frac{\partial \boldsymbol{k}_e}{\partial x_j} = \frac{\partial \boldsymbol{B}_e^T}{\partial x_j} D_e \boldsymbol{B}_e + \boldsymbol{B}_e^T \frac{\partial D_e}{\partial x_j}\boldsymbol{B}_e + \boldsymbol{B}_e^T D_e \frac{\partial \boldsymbol{B}_e}{\partial x_j}. \tag{6.14}$$

Let us calculate $\partial \boldsymbol{k}_e / \partial x_{pq}$, where x_{pq} are the coordinates of the end points of bar e, as shown in Fig. 6.2. Obviously, $\boldsymbol{k}_e$ is not affected by changes in any other coordinates. The coordinate system that was previously denoted the xy-system is now

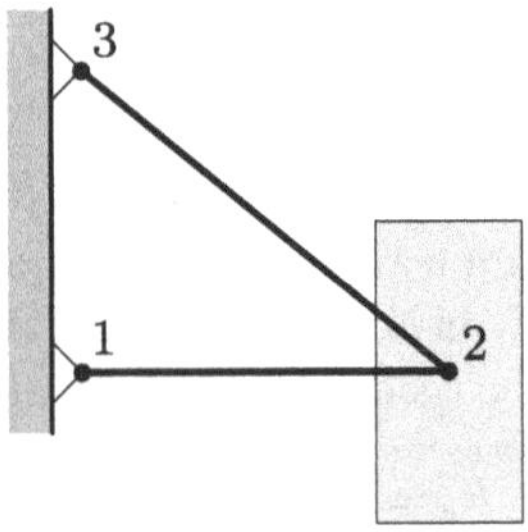

Fig. 6.1 Shape optimization of a two-bar truss

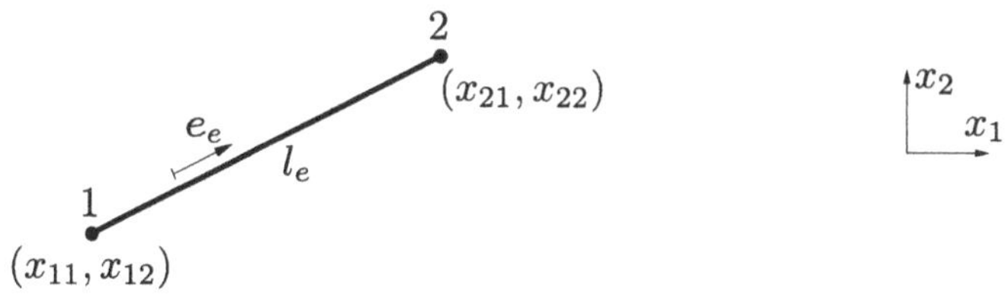

Fig. 6.2 A general bar e

called the x_1x_2-system. Coordinate x_{pq} is the x_q-coordinate of end point p (1 or 2). We introduce the column matrix $\boldsymbol{h}$ as

$$\boldsymbol{h} = \begin{cases} [-1\ 0]^T & \text{if } pq = 11 \\ [0\ -1]^T & \text{if } pq = 12 \\ [1\ 0]^T & \text{if } pq = 21 \\ [0\ 1]^T & \text{if } pq = 22. \end{cases}$$

We have for the unit vector $\boldsymbol{e}_e$ directed from end point 1 to 2 that

$$\boldsymbol{e}_e = \frac{1}{l_e} \begin{bmatrix} x_{21} - x_{11} \\ x_{22} - x_{12} \end{bmatrix},$$

where $l_e = \sqrt{(x_{21} - x_{11})^2 + (x_{22} - x_{12})^2}$. Differentiation of l_e gives

$$\frac{\partial l_e}{\partial x_{pq}} = \boldsymbol{h}^T \boldsymbol{e}_e. \tag{6.15}$$

The derivative of D_e becomes

$$\frac{\partial D_e}{\partial x_{pq}} = -\frac{E_e A_e}{l_e^2} \frac{\partial l_e}{\partial x_{pq}} = -\frac{E_e A_e}{l_e^2} \boldsymbol{h}^T \boldsymbol{e}_e.$$

If we differentiate $\boldsymbol{e}_e$, we obtain

$$\frac{\partial \boldsymbol{e}_e}{\partial x_{pq}} = -\frac{1}{l_e^2} \frac{\partial l_e}{\partial x_{pq}} \begin{bmatrix} x_{21} - x_{11} \\ x_{22} - x_{12} \end{bmatrix} + \frac{1}{l_e} \boldsymbol{h} = \frac{1}{l_e} (\boldsymbol{I} - \boldsymbol{e}_e \boldsymbol{e}_e^T) \boldsymbol{h},$$

where $\boldsymbol{I}$ is a 2-by-2 unit matrix. Since $\boldsymbol{B}_e = [-\boldsymbol{e}_e^T\ \boldsymbol{e}_e^T]$, $\partial \boldsymbol{B}_e / \partial x_{pq}$ immediately follows. Thus, all terms in (6.14) have been calculated.

If the applied load is design dependent we have to calculate $\partial \boldsymbol{f}_e^a / \partial x_{pq}$. For example, if gravity g acts in the negative x_2 direction, then

$$\boldsymbol{f}_e^a = \begin{bmatrix} 0 \\ -A_e l_e \rho g/2 \\ 0 \\ -A_e l_e \rho g/2 \end{bmatrix},$$

where ρ is the density. The sensitivity is obtained from (6.15) as

$$\frac{\partial \boldsymbol{f}_e^a}{\partial x_{pq}} = -\frac{A_e \rho g}{2} \begin{bmatrix} 0 \\ \boldsymbol{h}^T \boldsymbol{e}_e \\ 0 \\ \boldsymbol{h}^T \boldsymbol{e}_e \end{bmatrix}.$$

We end this discussion of sensitivity calculations for trusses by noting that we may also consider problems of *simultaneous* sizing and shape optimization, i.e. problems where both optimal cross-sectional areas of bars as well as optimal nodal positions should be determined.

6.3.2 Plane Sheets

A sheet is an example of a distributed parameter system since there is an infinite number of degrees-of-freedom. In order to be able to solve optimization problems involving arbitrary sheets, we discretize the state variables (displacement, stress, strain, etc.) using the finite element method. The design variables may also be infinite; e.g. one may want to determine the thickness of the sheet as a completely unknown continuous function of the location in the sheet. The design variables will also be discretized to obtain an optimization problem amenable to computer solution. If the optimal thickness distribution of the sheet is sought, one may for instance say that the thickness should be constant in each finite element. After discretization of the state variables and the design variables, we will have a discrete optimization problem that is similar to that of a naturally discrete system such as a truss. We will perform the sensitivity analysis on the finite dimensional, discretized problem. Although much more complicated, it is also possible to perform the sensitivity analysis for the original infinite-dimensional problem (possibly after a discretization of the design variables), and then discretize the sensitivity equations obtained. We refer to Haftka and Gürdal [18] for an introduction to infinite-dimensional sensitivity analysis, and Choi and Kim [12, 13] for a more detailed treatment.

We start by reviewing some fundamentals of finite element modeling of plane sheets, cf. e.g. Hughes [21]. All finite elements of the same type in the mesh are mapped from a single parent element by a one-to-one mapping $\boldsymbol{\xi} \mapsto \boldsymbol{x}(\boldsymbol{\xi})$, where $\boldsymbol{x} = [x_1\ x_2]^T$ and $\boldsymbol{\xi} = [\xi_1\ \xi_2]^T$, see Fig. 6.3. Illustrating with a four-nodded element, the parent element occupies the region $-1 \leq \xi_1 \leq 1,\ -1 \leq \xi_2 \leq 1$, denoted $\hat{\Omega}$, in the $\xi_1\xi_2$-system. The location Ω_e of the finite element in the x_1x_2-system will in general depend on the design variables. These will in the following be denoted α_i rather than x_i to avoid confusion between design variables and coordinates. Using an isoparametric formulation, the geometry and displacements are interpolated by the same shape functions:

$$\boldsymbol{x}(\boldsymbol{\xi}) = \boldsymbol{X}^T \boldsymbol{N}_v \tag{6.16}$$

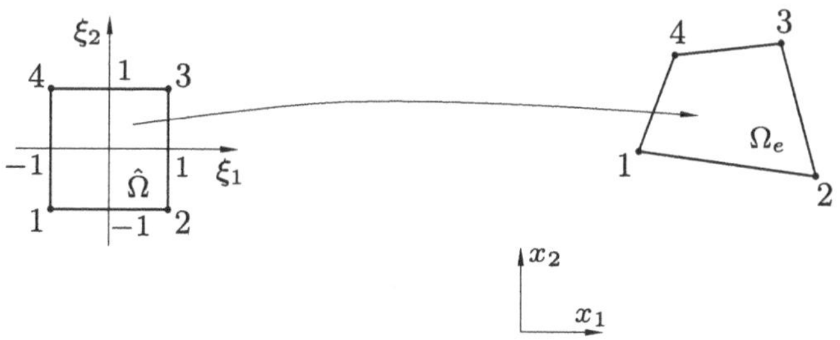

Fig. 6.3 Mapping of the parent element to the finite element e

$$\tilde{\boldsymbol{u}}(\boldsymbol{\xi}) = \boldsymbol{U}^T \boldsymbol{N}_v.$$

The matrix $\boldsymbol{X}$ holds the nodal coordinates of element e as

$$\boldsymbol{X} = \begin{bmatrix} x_{11} & x_{12} \\ \vdots & \vdots \\ x_{n_n 1} & x_{n_n 2} \end{bmatrix}, \tag{6.17}$$

where n_n denotes the number of nodes in the element, and x_{pq} is the x_q-coordinate of node p. The column matrix $\boldsymbol{N}_v$ contains the shape functions:

$$\boldsymbol{N}_v = \begin{bmatrix} N_1(\boldsymbol{\xi}) \\ \vdots \\ N_{n_n}(\boldsymbol{\xi}) \end{bmatrix}. \tag{6.18}$$

Shape function i is 1 at node i and zero for all other nodes. As an example, the shape functions for a four-nodded element are

$$N_1(\boldsymbol{\xi}) = \frac{1}{4}(1-\xi_1)(1-\xi_2)$$

$$N_2(\boldsymbol{\xi}) = \frac{1}{4}(1+\xi_1)(1-\xi_2)$$

$$N_3(\boldsymbol{\xi}) = \frac{1}{4}(1+\xi_1)(1+\xi_2)$$

$$N_4(\boldsymbol{\xi}) = \frac{1}{4}(1-\xi_1)(1+\xi_2).$$

The elements of the matrix $\boldsymbol{U}$,

$$\boldsymbol{U} = \begin{bmatrix} u_{11} & u_{12} \\ \vdots & \vdots \\ u_{n_n 1} & u_{n_n 2}, \end{bmatrix},$$

are the displacements at the nodes. The displacements at an arbitrary point $\boldsymbol{\xi}$ is denoted $\tilde{\boldsymbol{u}} = [\tilde{u}_1\ \tilde{u}_2]^T$. A tilde ($\tilde{}$) is used here since $\boldsymbol{u}$ is used for the global displacement column matrix that contains the displacements of all nonsuppressed nodes in the finite element mesh.

The element stiffness matrix may be shown to be, see e.g. Hughes [21],

$$\boldsymbol{k}_e = \int_{\hat{\Omega}} \boldsymbol{B}^T \boldsymbol{D} \boldsymbol{B} |\boldsymbol{J}| t \, d\hat{\Omega}, \tag{6.19}$$

where we have skipped index e for all terms in the integrand. This integral is obtained numerically by determining the value of the integrand at certain points, so-called Gauss points, in the element. The strain displacement matrix $\boldsymbol{B}$ is

$$\boldsymbol{B} = \begin{bmatrix} \dfrac{\partial N_1}{\partial x_1} & 0 & \cdots & \dfrac{\partial N_{n_n}}{\partial x_1} & 0 \\ 0 & \dfrac{\partial N_1}{\partial x_2} & \cdots & 0 & \dfrac{\partial N_{n_n}}{\partial x_2} \\ \dfrac{\partial N_1}{\partial x_2} & \dfrac{\partial N_1}{\partial x_1} & \cdots & \dfrac{\partial N_{n_n}}{\partial x_2} & \dfrac{\partial N_{n_n}}{\partial x_1} \end{bmatrix}. \tag{6.20}$$

For the case of plane stress, e.g., the stress strain matrix $\boldsymbol{D}$ is written

$$\boldsymbol{D} = \frac{E}{1-\nu^2} \begin{bmatrix} 1 & \nu & 0 \\ \nu & 1 & 0 \\ 0 & 0 & \dfrac{1-\nu}{2} \end{bmatrix},$$

where E is Young's modulus and ν is Poisson's ratio. The matrix

$$\boldsymbol{J} = \begin{bmatrix} \dfrac{\partial x_1}{\partial \xi_1} & \dfrac{\partial x_2}{\partial \xi_1} \\ \dfrac{\partial x_1}{\partial \xi_2} & \dfrac{\partial x_2}{\partial \xi_2} \end{bmatrix}$$

is the Jacobian of the mapping $\boldsymbol{\xi} \mapsto \boldsymbol{x}(\boldsymbol{\xi})$ and $|\boldsymbol{A}|$ denotes the determinant of an arbitrary matrix $\boldsymbol{A}$. Finally, t is the thickness of the element.

6.3.2.1 Sizing and Topology Optimization

Let us assume that we approximate the thickness of the sheet as constant in each element e. For sizing and topology optimization, we then choose $\alpha_e = t_e$, $e = 1, \ldots, n_e$, where n_e is the number of elements in the mesh. Thus, t_e may be moved out of the integral in (6.19):

$$\boldsymbol{k}_e = t_e \int_{\hat{\Omega}} \boldsymbol{B}^T \boldsymbol{D} \boldsymbol{B} |\boldsymbol{J}| \, d\hat{\Omega} = t_e \boldsymbol{k}_e^0,$$

with an obvious definition of $\boldsymbol{k}_e^0$. Consequently, the sensitivity of the element stiffness matrix becomes simply

$$\frac{\partial \boldsymbol{k}_e}{\partial \alpha_i} = \begin{cases} \boldsymbol{k}_e^0 & \text{if } e = i \\ \boldsymbol{0} & \text{otherwise.} \end{cases}$$

6.3.2.2 Shape Optimization

In shape optimization we assume that the shape of some boundary curves is controlled by a number of design variables α_i, $i = 1, 2, \ldots$, see Fig. 6.4. In Chap. 7 we will present a detailed account of shape representation of boundary curves. In shape optimization the region Ω_e will depend on the design: $\Omega_e = \Omega_e(\alpha_i)$, but the region $\hat{\Omega}$ is fixed, as always. Since the thickness t and the matrix $\boldsymbol{D}$ are constant, differentiation of (6.19) gives

$$\frac{\partial \boldsymbol{k}_e}{\partial \alpha_i} = \int_{\hat{\Omega}} \left(\frac{\partial \boldsymbol{B}^T}{\partial \alpha_i} \boldsymbol{D}\boldsymbol{B}|\boldsymbol{J}| + \boldsymbol{B}^T \boldsymbol{D} \frac{\partial \boldsymbol{B}}{\partial \alpha_i} |\boldsymbol{J}| + \boldsymbol{B}^T \boldsymbol{D}\boldsymbol{B} \frac{\partial |\boldsymbol{J}|}{\partial \alpha_i} \right) t \, d\hat{\Omega}. \tag{6.21}$$

Note that t may very well vary within the element, but it does not change because of changes in the design variables. Our task is now to find analytical formulas for $\partial \boldsymbol{B}/\partial \alpha_i$ and $\partial |\boldsymbol{J}|/\partial \alpha_i$. To the best of our knowledge, such formulas were first derived by Brockman [9]. Our presentation follows the neater derivation in Haslinger and Mäkinen [19].

First, we define two additional matrices:

$$\boldsymbol{G} = \begin{bmatrix} \dfrac{\partial N_1}{\partial x_1} & \cdots & \dfrac{\partial N_{n_n}}{\partial x_1} \\ \dfrac{\partial N_1}{\partial x_2} & \cdots & \dfrac{\partial N_{n_n}}{\partial x_2} \end{bmatrix}, \tag{6.22}$$

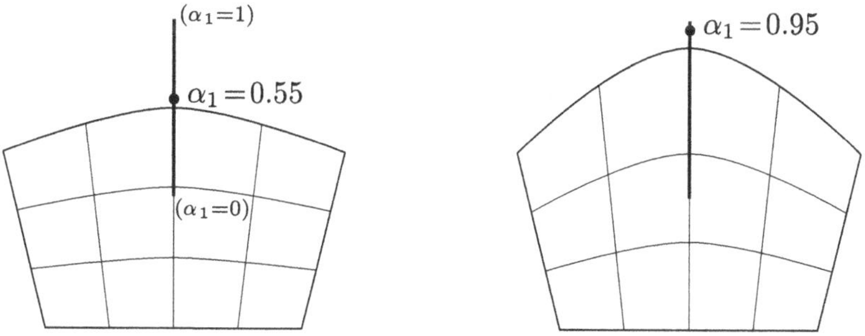

Fig. 6.4 Shape optimization of a sheet. Design variables determine the location of control points which in turn determine the shape of the boundary curve

and

$$\hat{\boldsymbol{G}} = \begin{bmatrix} \dfrac{\partial N_1}{\partial \xi_1} & \cdots & \dfrac{\partial N_{n_n}}{\partial \xi_1} \\ \dfrac{\partial N_1}{\partial \xi_2} & \cdots & \dfrac{\partial N_{n_n}}{\partial \xi_2} \end{bmatrix}. \tag{6.23}$$

Now, by the chain rule we have

$$\hat{\boldsymbol{G}} = \boldsymbol{J}\boldsymbol{G}. \tag{6.24}$$

Equation (6.16) gives

$$\boldsymbol{J} = \hat{\boldsymbol{G}}\boldsymbol{X}. \tag{6.25}$$

Clearly, $\hat{\boldsymbol{G}}$ is independent of the design ($\hat{\boldsymbol{G}}$ will be a function of $\boldsymbol{\xi}$, not $\boldsymbol{x}(\boldsymbol{\xi})$). For notational convenience, let us denote $\partial/\partial\alpha_i$ by a prime ($'$). We then have

$$\boldsymbol{0} = \hat{\boldsymbol{G}}' = \boldsymbol{J}'\boldsymbol{G} + \boldsymbol{J}\boldsymbol{G}'. \tag{6.26}$$

Also, (6.25) gives

$$\boldsymbol{J}' = \hat{\boldsymbol{G}}\boldsymbol{X}'. \tag{6.27}$$

From (6.26), (6.27) and (6.24) we get

$$\boldsymbol{G}' = -\boldsymbol{J}^{-1}\boldsymbol{J}'\boldsymbol{G} = -\boldsymbol{J}^{-1}\hat{\boldsymbol{G}}\boldsymbol{X}'\boldsymbol{G} = -\boldsymbol{G}\boldsymbol{X}'\boldsymbol{G}.$$

Next, we try to find $|\boldsymbol{J}|'$. We will make use of a simple result from linear algebra; for any nonsingular matrix $\boldsymbol{A}$ it holds that

$$|\boldsymbol{A}|' = |\boldsymbol{A}|\,\mathrm{tr}(\boldsymbol{A}^{-1}\boldsymbol{A}'), \tag{6.28}$$

where $\mathrm{tr}\,\boldsymbol{C}$ denotes the *trace* of an arbitrary square matrix $\boldsymbol{C}$, i.e. the sum of all diagonal terms of $\boldsymbol{C}$: $\mathrm{tr}\,\boldsymbol{C} = C_{11} + C_{22} + C_{33} + \cdots$. For a 2-by-2 matrix, such as $\boldsymbol{J}$, this result is most easily proven by direct calculation. Let us therefore study an arbitrary nonsingular matrix

$$\boldsymbol{A} = \begin{bmatrix} a_{11} & a_{12} \\ a_{21} & a_{22} \end{bmatrix}.$$

The determinant of $\boldsymbol{A}$ is

$$|\boldsymbol{A}| = a_{11}a_{22} - a_{12}a_{21},$$

which after differentiation gives

$$|\boldsymbol{A}|' = a_{11}'a_{22} + a_{11}a_{22}' - a_{12}'a_{21} - a_{12}a_{21}'. \tag{6.29}$$

The term $\boldsymbol{A}^{-1}\boldsymbol{A}'$ becomes

$$\begin{aligned}\boldsymbol{A}^{-1}\boldsymbol{A}' &= \frac{1}{|\boldsymbol{A}|}\begin{bmatrix} a_{22} & -a_{12} \\ -a_{21} & a_{11}\end{bmatrix}\begin{bmatrix} a'_{11} & a'_{12} \\ a'_{21} & a'_{22}\end{bmatrix} \\ &= \frac{1}{|\boldsymbol{A}|}\begin{bmatrix} a'_{11}a_{22} - a'_{21}a_{12} & a'_{12}a_{22} - a'_{22}a_{12} \\ a'_{21}a_{11} - a'_{11}a_{21} & a'_{22}a_{11} - a'_{12}a_{21}\end{bmatrix}.\end{aligned}$$

By adding the diagonal terms, multiplying with $|\boldsymbol{A}|$ and comparing with (6.29), we conclude that (6.28) holds.

Applying (6.28), and using (6.27) and (6.24), we have

$$|\boldsymbol{J}|' = |\boldsymbol{J}|\,\mathrm{tr}\big(\boldsymbol{J}^{-1}\boldsymbol{J}'\big) = |\boldsymbol{J}|\,\mathrm{tr}\big(\boldsymbol{J}^{-1}\hat{\boldsymbol{G}}\boldsymbol{X}'\big) = |\boldsymbol{J}|\,\mathrm{tr}(\boldsymbol{G}\boldsymbol{X}').$$

To summarize, we have shown that

$$\frac{\partial \boldsymbol{G}}{\partial \alpha_i} = -\boldsymbol{G}\frac{\partial \boldsymbol{X}}{\partial \alpha_i}\boldsymbol{G} \tag{6.30}$$

$$\frac{\partial |\boldsymbol{J}|}{\partial \alpha_i} = |\boldsymbol{J}|\,\mathrm{tr}\left(\boldsymbol{G}\frac{\partial \boldsymbol{X}}{\partial \alpha_i}\right). \tag{6.31}$$

Although derived for two-dimensional problems here, it is straightforward to see that these formulas are valid in the same form for one- and three-dimensional problems as well.

Once the sensitivity of $\boldsymbol{G}$ has been obtained, the sensitivity of $\boldsymbol{B}$ follows directly, cf. (6.22) and (6.20). Note that in $\partial \boldsymbol{B}/\partial \alpha_i$ only first order derivatives of the shape functions appear. These terms are exactly the same that are used in a standard finite element analysis when forming the equilibrium equations. Intuitively, one would think that second order derivatives of the shape functions would be needed in order to find the sensitivity of $\boldsymbol{B}$, since we should calculate $(\partial/\partial\alpha_i)(\partial N_j/\partial x_k)$ where the geometry (x_1 and x_2) depends on α_i. We still need to find $\partial \boldsymbol{X}/\partial \alpha_i$, i.e. the partial derivatives of the nodal positions with respect to the design variables. These derivatives will depend on how the shape is represented. In the next chapter we will see how they can be obtained.

Finally, we will discuss how design dependent loads may be included in the analysis. First, the following matrix with the shape functions is introduced:

$$\boldsymbol{N} = \begin{bmatrix} N_1 & 0 & \cdots & N_{n_n} & 0 \\ 0 & N_1 & \cdots & 0 & N_{n_n}\end{bmatrix}.$$

For simplicity, we will not treat design dependent tractions. For the body force $\boldsymbol{b}$ (force per unit area), the applied force on element e may be written

$$\boldsymbol{f}_e^a = \int_{\hat{\Omega}} \boldsymbol{N}^T\boldsymbol{b}|\boldsymbol{J}|t\,d\hat{\Omega}. \tag{6.32}$$

As examples of design dependent body forces, we may take gravity in the negative x_2-direction, and centrifugal forces due to rotation around the x_2-axis: $\boldsymbol{b} =$

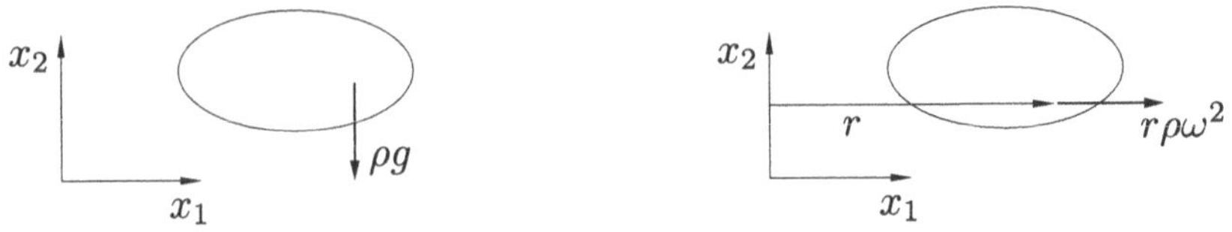

Fig. 6.5 Design dependent loads

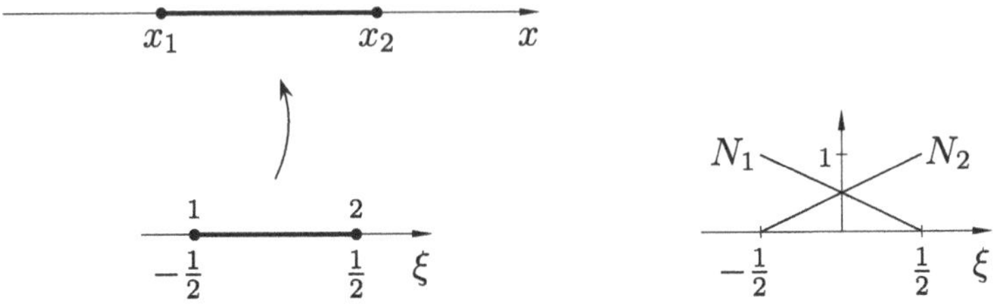

Fig. 6.6 Bar element

$[0\ -\rho g]^T$ and $\boldsymbol{b} = [r\rho\omega^2\ 0]^T$, where ρ is the density (mass per unit area), g is the acceleration due to gravity, ω is the angular speed, and the radius r may be interpolated as $r = \sum_{i=1}^{n_n} N_i x_{i1}$; see Fig. 6.5. If (6.32) is differentiated we obtain

$$\frac{\partial \boldsymbol{f}_e^a}{\partial \alpha_i} = \int_{\hat{\Omega}} \boldsymbol{N}^T \left(\frac{\partial \boldsymbol{b}}{\partial \alpha_i} |\boldsymbol{J}| + \boldsymbol{b} \frac{\partial |\boldsymbol{J}|}{\partial \alpha_i} \right) t\, d\hat{\Omega}. \tag{6.33}$$

Example 6.3 Consider a bar element where the coordinates of the end points x_1 and $x_2 > x_1$ are the design variables, see Fig. 6.6. The cross-sectional area is A, and Young's modulus is E. We would like to calculate the sensitivity of the element stiffness matrix, i.e. $\partial \boldsymbol{k}_e/\partial x_1$ and $\partial \boldsymbol{k}_e/\partial x_2$, by making use of the formulas derived for a sheet in this section. Since we in this problem deal with a one-dimensional bar rather than a two-dimensional sheet, the equations above need to be modified in a straightforward way.

The end points are collected in the column matrix $\boldsymbol{X}$, cf. (6.17), and the shape functions are collected in the column matrix $\boldsymbol{N}_v$, cf. (6.18):

$$\boldsymbol{X} = \begin{bmatrix} x_1 \\ x_2 \end{bmatrix}, \qquad \boldsymbol{N}_v = \begin{bmatrix} N_1(\xi) \\ N_2(\xi) \end{bmatrix} = \begin{bmatrix} \frac{1}{2} - \xi \\ \frac{1}{2} + \xi \end{bmatrix}.$$

The geometry is interpolated according to (6.16) as

$$x = \boldsymbol{X}^T \boldsymbol{N}_v = \left(\frac{1}{2} - \xi \right) x_1 + \left(\frac{1}{2} + \xi \right) x_2.$$

The matrix $\hat{\boldsymbol{G}}$ in (6.23) becomes

$$\hat{\boldsymbol{G}} = \begin{bmatrix} \frac{\partial N_1}{\partial \xi} & \frac{\partial N_2}{\partial \xi} \end{bmatrix} = [-1\ \ 1],$$

whereas the matrix $\boldsymbol{G}$ in (6.22) may be written as

$$\boldsymbol{G} = \begin{bmatrix} \frac{\partial N_1}{\partial x} & \frac{\partial N_2}{\partial x} \end{bmatrix}.$$

Here,

$$\frac{\partial N_1}{\partial x} = \frac{\partial N_1}{\partial \xi}\frac{\partial \xi}{\partial x} = \frac{\partial N_1}{\partial \xi}\frac{1}{\frac{dx}{d\xi}} = -\frac{1}{x_2 - x_1},$$

and similarly

$$\frac{\partial N_1}{\partial x} = \frac{1}{x_2 - x_1}.$$

The strain-displacement matrix $\boldsymbol{B}$ is

$$\boldsymbol{B} = \begin{bmatrix} \frac{\partial N_1}{\partial x} & \frac{\partial N_2}{\partial x} \end{bmatrix} = \frac{1}{x_2 - x_1}\begin{bmatrix} -1 & 1 \end{bmatrix}.$$

Evidently, for this problem, $\boldsymbol{B} = \boldsymbol{G}$. Note that $\boldsymbol{B}$ relates the strain ε of the bar to the displacements $\boldsymbol{u}$ of the end points: $\varepsilon = \boldsymbol{B}\boldsymbol{u}$. Thus, the $\boldsymbol{B}$-matrix used here has a different interpretation than the $\boldsymbol{B}$-matrices used for trusses previously where $\boldsymbol{B}$ is defined to relate the elongation δ to the displacements: $\delta = \boldsymbol{B}\boldsymbol{u}$.

The Jacobian $\boldsymbol{J}$ is in this case a scalar:

$$\boldsymbol{J} = J = \frac{\partial x}{\partial \xi} = x_2 - x_1.$$

The element stiffness matrix reads

$$\boldsymbol{k}_e = \int_{-1/2}^{1/2} \boldsymbol{B}^T \boldsymbol{D}\boldsymbol{B}|\boldsymbol{J}|A\, d\xi,$$

where the stress strain matrix $\boldsymbol{D}$ in this case will be the Young's modulus E. The sensitivities of $\boldsymbol{k}_e$ are obtained as

$$\frac{\partial \boldsymbol{k}_e}{\partial x_i} = AE\int_{-1/2}^{1/2}\left(J\frac{\partial \boldsymbol{G}^T}{\partial x_i}\boldsymbol{G} + J\boldsymbol{G}^T\frac{\partial \boldsymbol{G}}{\partial x_i} + \frac{\partial J}{\partial x_i}\boldsymbol{G}^T\boldsymbol{G}\right) d\xi. \tag{6.34}$$

The sensitivities of $\boldsymbol{G}$ and J may be obtained by direct differentiation, but we will make use of the general formulas (6.30) and (6.31):

$$\frac{\partial \boldsymbol{G}}{\partial x_1} = -\boldsymbol{G}\frac{\partial \boldsymbol{X}}{\partial x_1}\boldsymbol{G} = -\frac{1}{(x_2 - x_1)^2}\begin{bmatrix} -1 & 1 \end{bmatrix}\begin{bmatrix} 1 \\ 0 \end{bmatrix}\begin{bmatrix} -1 & 1 \end{bmatrix} = -\frac{1}{(x_2 - x_1)^2}\begin{bmatrix} 1 & -1 \end{bmatrix}$$

$$\frac{\partial J}{\partial x_1} = J\,\mathrm{tr}\left(\boldsymbol{G}\frac{\partial \boldsymbol{X}}{\partial x_1}\right) = (x_2 - x_1)\,\mathrm{tr}\left(\frac{1}{x_2 - x_1}\begin{bmatrix} -1 & 1 \end{bmatrix}\begin{bmatrix} 1 \\ 0 \end{bmatrix}\right) = -1$$

$$\frac{\partial \boldsymbol{G}}{\partial x_2} = \frac{1}{(x_2 - x_1)^2} \begin{bmatrix} 1 & -1 \end{bmatrix}$$

$$\frac{\partial J}{\partial x_2} = 1.$$

Insertion of these expressions into (6.34) gives us the sensitivities of $\boldsymbol{k}_e$ as

$$\frac{\partial \boldsymbol{k}_e}{\partial x_1} = \frac{AE}{(x_2 - x_1)^2} \begin{bmatrix} 1 & -1 \\ -1 & 1 \end{bmatrix}$$

$$\frac{\partial \boldsymbol{k}_e}{\partial x_2} = -\frac{AE}{(x_2 - x_1)^2} \begin{bmatrix} 1 & -1 \\ -1 & 1 \end{bmatrix}.$$

For this simple example, the sensitivities of $\boldsymbol{k}_e$ could of course have been obtained easier by differentiating $\boldsymbol{k}_e$ directly, where

$$\boldsymbol{k}_e = \frac{AE}{x_2 - x_1} \begin{bmatrix} 1 & -1 \\ -1 & 1 \end{bmatrix}.$$

6.4 Exercises

Exercise 6.1 Consider the function $f(\boldsymbol{x}) = \boldsymbol{u}(\boldsymbol{x})^T \boldsymbol{u}(\boldsymbol{x})$, where $\boldsymbol{u}(\boldsymbol{x})$ is an implicit function of $\boldsymbol{x}$ through the equilibrium equations $\boldsymbol{K}(\boldsymbol{x})\boldsymbol{u}(\boldsymbol{x}) = \boldsymbol{F}(\boldsymbol{x})$. The sensitivities $\partial \boldsymbol{K}(\boldsymbol{x})/\partial x_j$ and $\partial \boldsymbol{F}(\boldsymbol{x})/\partial x_j$ are known. Calculate the sensitivities $\partial f(\boldsymbol{x})/\partial x_j$

a) by the direct method,
b) by the adjoint method.

Exercise 6.2 A two-bar truss should be optimized, see Fig. 6.7. The x-coordinate of the free node, x, as well as the cross-sectional areas, A_1 and A_2, are taken as design variables. Calculate the sensitivities of the global stiffness matrix $\boldsymbol{K}$, i.e.

$$\frac{\partial \boldsymbol{K}}{\partial A_1}, \quad \frac{\partial \boldsymbol{K}}{\partial A_2}, \quad \frac{\partial \boldsymbol{K}}{\partial x}.$$

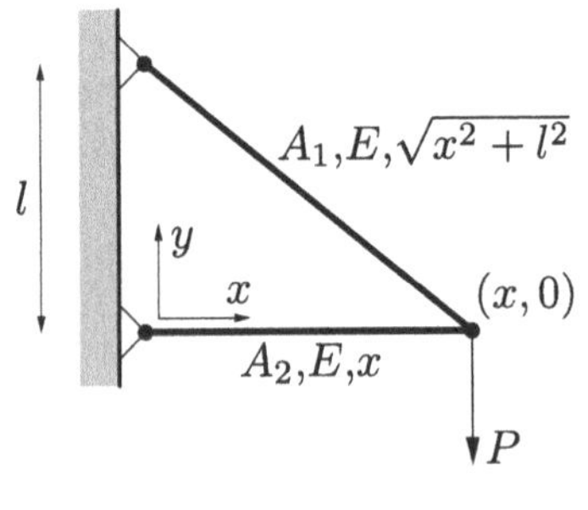

Fig. 6.7 The two-bar truss of Exercise 6.2

Fig. 6.8 The bar of Exercise 6.3

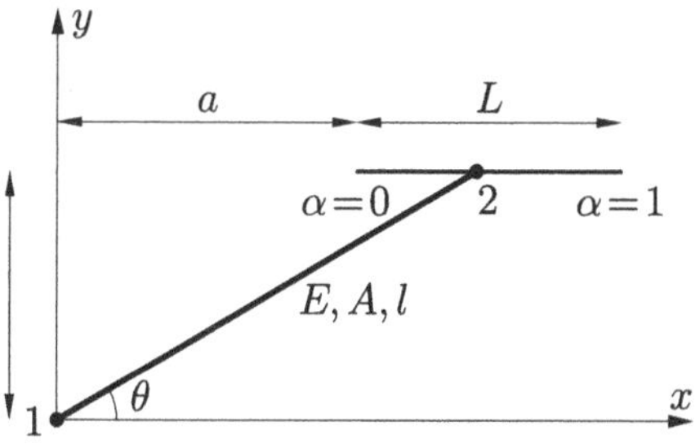

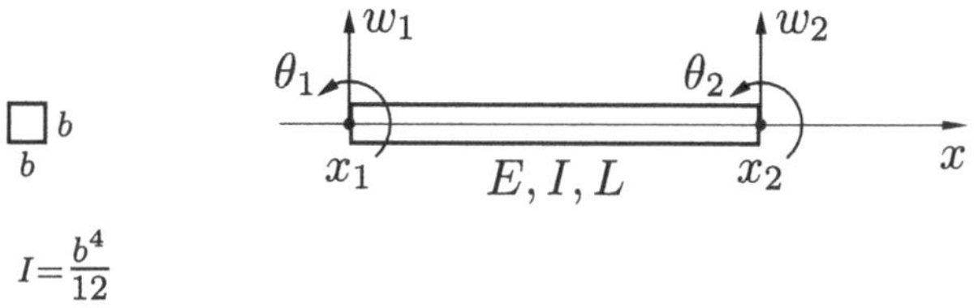

Fig. 6.9 The beam of Exercise 6.4

Exercise 6.3 One of the end points of a bar is fixed, whereas the other end point, point 2, is allowed to move along a fixed horizontal line of length L, see Fig. 6.8. The leftmost point on this line has the coordinates (a, b). The location of point 2 is controlled by a design variable α which changes linearly from 0 to 1 on the line. The cross-sectional area A of the bar is also allowed to change, whereas Young's modulus E is constant. The stiffness matrix of the bar is given in (6.13). Calculate the sensitivities $\partial \boldsymbol{k}/\partial \alpha$ and $\partial \boldsymbol{k}/\partial A$.

Exercise 6.4 For the beam element in Fig. 6.9, with degrees-of-freedom $[w_1\ \theta_1\ w_2\ \theta_2]^T$, the stiffness matrix may be written as

$$\boldsymbol{k}_e = \frac{EI}{L^3}\begin{bmatrix} 12 & 6L & -12 & 6L \\ 6L & 4L^2 & -6L & 2L^2 \\ -12 & -6L & 12 & -6L \\ 6L & 2L^2 & -6L & 4L^2 \end{bmatrix},$$

where L is the length of the beam, E is Young's modulus and I is the moment of inertia of the cross section of the beam. The end points x_1 and x_2 as well as the cross-sectional area A are taken as design variables. The cross section is square with dimensions $b \times b$. Calculate the sensitivities

$$\frac{\partial \boldsymbol{k}_e}{\partial x_1}, \qquad \frac{\partial \boldsymbol{k}_e}{\partial x_2}, \qquad \frac{\partial \boldsymbol{k}_e}{\partial A}.$$

Exercise 6.5 A bar-element is subjected to a centrifugal load $b = \rho x \omega^2$, where ρ is the mass per unit length, and ω is the angular velocity of the bar around the y-axis, see Fig. 6.10. The isoparametric element has two nodes with coordinates x_1

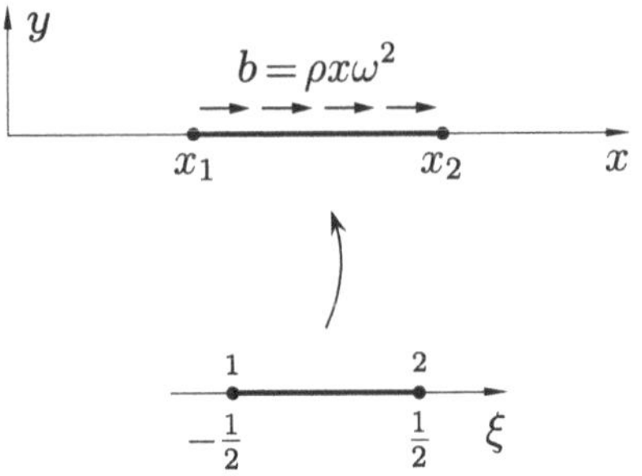

Fig. 6.10 The one-dimensional bar element of Exercise 6.5

and x_2. The shape functions of the parent element are $N_1(\xi) = 1/2 - \xi$ and $N_2(\xi) = 1/2 + \xi$. The nodal coordinates x_1 and x_2 are used as design variables. Calculate the sensitivities of the element force vector $\boldsymbol{f}_e^a$, i.e. $\partial \boldsymbol{f}_e^a/\partial x_1$ and $\partial \boldsymbol{f}_e^a/\partial x_2$. Do this in two ways: 1) form $\boldsymbol{f}_e^a$ and then differentiate the expression obtained; 2) use (6.33).

Exercise 6.6 Study a function $g_i(\boldsymbol{x},\boldsymbol{u})$, where $\boldsymbol{u}$ is an implicit function of $\boldsymbol{x}$ defined through the equilibrium equations $\boldsymbol{K}(\boldsymbol{x})\boldsymbol{u} = \boldsymbol{F}(\boldsymbol{x})$. In (6.8), the first order derivatives of g_i are obtained using the adjoint method as

$$\frac{dg_i(\boldsymbol{x},\boldsymbol{u}(\boldsymbol{x}))}{dx_j} = \frac{\partial g_i(\boldsymbol{x},\boldsymbol{u})}{\partial x_j} + \boldsymbol{\lambda}_i^T\left(\frac{\partial \boldsymbol{F}(\boldsymbol{x})}{\partial x_j} - \frac{\partial \boldsymbol{K}(\boldsymbol{x})}{\partial x_j}\boldsymbol{u}(\boldsymbol{x})\right),$$

where $\boldsymbol{\lambda}_i$ is obtained from the equation

$$\boldsymbol{K}(\boldsymbol{x})\boldsymbol{\lambda}_i = \left(\frac{\partial g_i(\boldsymbol{x},\boldsymbol{u})}{\partial \boldsymbol{u}}\right)^T, \tag{6.35}$$

and $\partial g_i/\partial \boldsymbol{u}$ is defined to be a row matrix. It proves convenient in this exercise to write $dg_i(\boldsymbol{x},\boldsymbol{u}(\boldsymbol{x}))/dx_j$ for the complete x_j-derivative of g_i instead of $\partial \hat{g}_i(\boldsymbol{x})/\partial x_j$ as we have done previously.

Show that the second order derivatives of g_i may be written as

$$\begin{aligned}\frac{d^2 g_i}{dx_j dx_k} &= \frac{\partial^2 g_i}{\partial x_j \partial x_k} + \frac{\partial^2 g_i}{\partial \boldsymbol{u} \partial x_j}\frac{\partial \boldsymbol{u}}{\partial x_k} + \frac{\partial^2 g_i}{\partial \boldsymbol{u} \partial x_k}\frac{\partial \boldsymbol{u}}{\partial x_j} + \left(\frac{\partial \boldsymbol{u}}{\partial x_k}\right)^T \frac{\partial^2 g_i}{\partial \boldsymbol{u}^2}\frac{\partial \boldsymbol{u}}{\partial x_j}\\ &+ \boldsymbol{\lambda}_i^T\left(\frac{\partial^2 \boldsymbol{F}}{\partial x_j \partial x_k} - \frac{\partial^2 \boldsymbol{K}}{\partial x_j \partial x_k}\boldsymbol{u} - \frac{\partial \boldsymbol{K}}{\partial x_j}\frac{\partial \boldsymbol{u}}{\partial x_k} - \frac{\partial \boldsymbol{K}}{\partial x_k}\frac{\partial \boldsymbol{u}}{\partial x_j}\right),\end{aligned}$$

where $\boldsymbol{\lambda}_i$ is obtained from (6.35), $\partial^2 g_i/\partial \boldsymbol{u}^2$ is a matrix,

$$\frac{\partial \boldsymbol{u}(\boldsymbol{x})}{\partial x_j} = \boldsymbol{K}(\boldsymbol{x})^{-1}\left(\frac{\partial \boldsymbol{F}(\boldsymbol{x})}{\partial x_j} - \frac{\partial \boldsymbol{K}(\boldsymbol{x})}{\partial x_j}\boldsymbol{u}(\boldsymbol{x})\right),$$

and similarly for $\partial \boldsymbol{u}/\partial x_k$.

Exercise 6.7 A sensitivity analysis should be performed for the two-bar truss in Fig. 6.11 by calculating the derivatives $\partial r/\partial A_1$ and $\partial r/\partial A_2$, where $r = -u_y$ is the

Fig. 6.11 The two-bar truss of Exercise 6.7

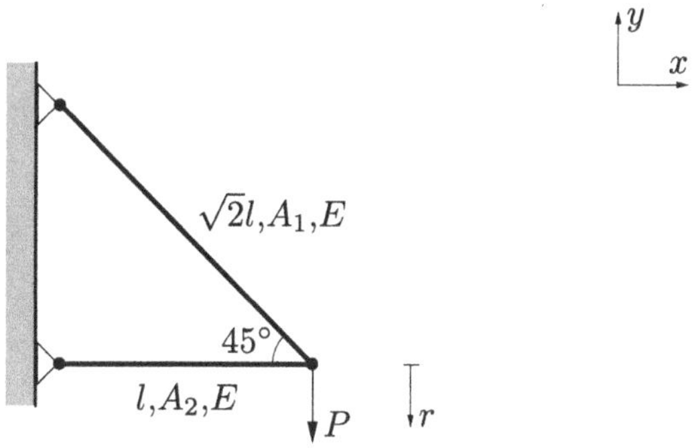

downward displacement of the free node, and A_i is the cross-sectional area of bar i. It may be shown that the equilibrium equations read

$$\begin{bmatrix} 0 \\ -P \end{bmatrix} = \frac{E}{2\sqrt{2}l} \begin{bmatrix} A_1 + 2\sqrt{2}A_2 & -A_1 \\ -A_1 & A_1 \end{bmatrix} \begin{bmatrix} u_x \\ u_y \end{bmatrix}.$$

It holds that $P = 1$ kN, $E = 210$ GPa, $l = 0.1$ m.

Calculate the sensitivities of r at $(A_1, A_2) = (10^{-4}, 10^{-4})$ m^2 in four different ways using:

a) The direct method.

b) The adjoint method.

c) The numerical method for a forward difference approximation. Use different design steps between $h = 10^{-20}$ m^2 and $h = 10^{-1}$ m^2, and plot the sensitivity as a function of h.

d) The numerical method for a central difference approximation. Plot the sensitivity as a function of h.

Chapter 7
Two-Dimensional Shape Optimization

In order to optimize the shape of a structure, one naturally has to be able to control the shape of its boundary using some design variables. In Sect. 6.3.2, the sensitivity analysis for shape optimization of sheets was described. It was concluded that the nodal sensitivities, i.e. the partial derivatives of the nodal positions with respect to the design variables were needed. These sensitivities will depend on how the shape is represented and also on how the finite element mesh is generated. In this chapter we will discuss how the nodal sensitivities may be calculated for two-dimensional structures such as plane sheets or axisymmetric bodies.

7.1 Shape Representation

A natural idea that springs to mind when figuring out how to optimize the shape of a sheet is to use the coordinates of the finite element nodes as design variables, just as is done for trusses. It turns out, however, that with this choice the boundaries tend to become extremely jagged, leading to inaccurate stress calculations. Alternatively, the boundaries can be described using polynomial functions and using design variables that control their shape. In order to represent complex geometries one may use either high-order polynomials or use low-order piecewise polynomials functions to form *spline curves*. When high-order polynomials are used, experience has shown that the boundaries may become highly oscillatory. Good results are obtained if instead low-order splines represent the boundary. Examples include Bézier splines, B-splines and nonuniform rational B-splines (NURBS).

The shape of a spline may be controlled by a number of control vertices $\boldsymbol{V}_i$, $i = 0, 1, \ldots,$ see Fig. 7.1. When these vertices move, the shape of the curve changes. In order to perform a shape optimization we therefore introduce design variables α_i to those control vertices that are allowed to move during the iteration process. Figure 7.2(a) depicts a control vertex $\boldsymbol{V}_i = [x_i \; y_i]^T$ that is allowed to move along a straight line, a so-called *span*. The end points of the span, $\boldsymbol{V}_i^{\min} = [x_i^{\min} \; y_i^{\min}]^T$ and $\boldsymbol{V}_i^{\max} = [x_i^{\max} \; y_i^{\max}]^T$, are assumed given. Along each span, a design variable α_i varies linearly. At $\boldsymbol{V}_i^{\min}$, $\alpha_i = 0$, and at $\boldsymbol{V}_i^{\max}$, $\alpha_i = 1$. Thus, the location of the control vertex $\boldsymbol{V}_i$ is given by

$$\boldsymbol{V}_i = \boldsymbol{V}_i^{\min} + \alpha_i \boldsymbol{L}_i, \quad 0 \leq \alpha_i \leq 1, \tag{7.1}$$

where the constant vector $\boldsymbol{L}_i$ is defined as $\boldsymbol{L}_i = \boldsymbol{V}_i^{\max} - \boldsymbol{V}_i^{\min}$. Differentiation of (7.1) yields the sensitivity of the control vertex as

$$\frac{\partial \boldsymbol{V}_i}{\partial \alpha_i} = \boldsymbol{L}_i. \tag{7.2}$$

P.W. Christensen, A. Klarbring, *An Introduction to Structural Optimization*,

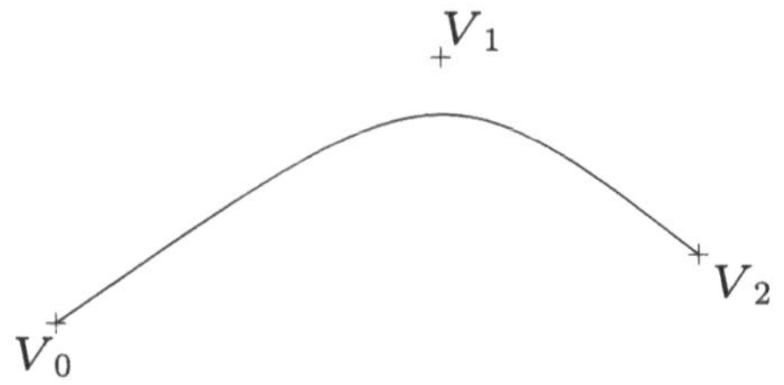

Fig. 7.1 Control vertices of a spline curve

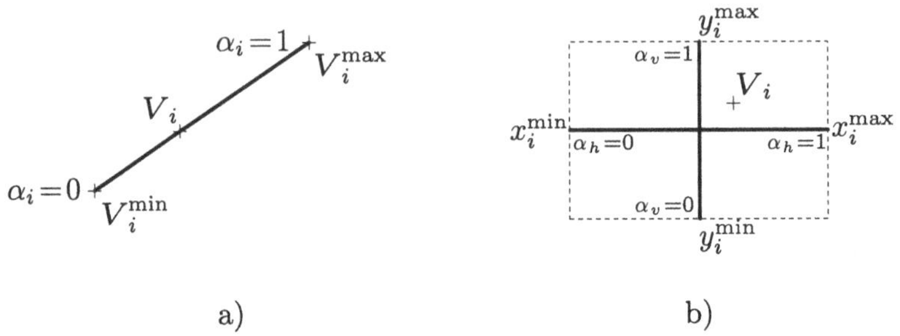

Fig. 7.2 Control vertices including spans

By making use of *two* spans for a certain control vertex, one horizontal and the other vertical, that vertex will be able to move inside a box, as illustrated in Fig. 7.2(b).

In what follows, we will present some basic properties of Bézier and B-splines. For a detailed account of these curves, including the more complex NURBS, see Piegl and Tiller [28] and Rogers [29].

7.1.1 Bézier Splines

An nth degree Bézier spline is defined as

$$\boldsymbol{r}(u) = \sum_{i=0}^{n} B_{i,n}(u)\boldsymbol{V}_i, \quad 0 \le u \le 1, \tag{7.3}$$

where $\boldsymbol{r}(u) = [x(u)\ y(u)]^T$ and the nth degree *Bernstein polynomials* $B_{i,n}$ are defined recursively as

$$B_{i,n}(u) = (1-u)B_{i,n-1}(u) + uB_{i-1,n-1}(u), \tag{7.4}$$

where $B_{0,0}(u) = 1$ and $B_{i,n}(u) = 0$ for $i < 0$ and $i > n$. Thus, u is a scalar variable that uniquely determines a point on the spline. It may be shown that the $B_{i,n}$:s take the explicit form

$$B_{i,n}(u) = \frac{n!}{i!(n-i)!}u^i(1-u)^{n-i}.$$

It holds that $\boldsymbol{r}(0) = \boldsymbol{V}_0$ and $\boldsymbol{r}(1) = \boldsymbol{V}_n$, i.e. $\boldsymbol{V}_0$ and $\boldsymbol{V}_n$ are the start and end points of the curve, respectively. Differentiation of (7.3) with respect to u leads after some calculations to

$$\frac{d\boldsymbol{r}(0)}{du} = n(\boldsymbol{V}_1 - \boldsymbol{V}_0) \tag{7.5}$$

$$\frac{d\boldsymbol{r}(1)}{du} = n(\boldsymbol{V}_n - \boldsymbol{V}_{n-1}). \tag{7.6}$$

Thus, the tangent of the curve at the start (end) point only depends on the location of the first (last) two control vertices.

For $n = 1$, the Bernstein polynomials are

$$B_{0,1} = (1-u)B_{0,0} + uB_{-1,0} = 1-u$$
$$B_{1,1} = (1-u)B_{1,0} + uB_{0,0} = u,$$

so that the first degree Bézier spline becomes

$$\boldsymbol{r}(u) = (1-u)\boldsymbol{V}_0 + u\boldsymbol{V}_1,$$

i.e. a straight line between the points $\boldsymbol{V}_0$ and $\boldsymbol{V}_1$. Putting $n = 2$, we have

$$B_{0,2} = (1-u)B_{0,1} + uB_{-1,1} = (1-u)^2$$
$$B_{1,2} = (1-u)B_{1,1} + uB_{0,1} = 2u(1-u)$$
$$B_{2,2} = (1-u)B_{2,1} + uB_{1,1} = u^2,$$

which gives the second degree curve

$$\boldsymbol{r}(u) = (1-u)^2\boldsymbol{V}_0 + 2u(1-u)\boldsymbol{V}_1 + u^2\boldsymbol{V}_2. \tag{7.7}$$

From

$$B_{0,3} = (1-u)B_{0,2} + uB_{-1,2} = (1-u)^3$$
$$B_{1,3} = (1-u)B_{1,2} + uB_{0,2} = 3u(1-u)^2$$
$$B_{2,3} = (1-u)B_{2,2} + uB_{1,2} = 3u^2(1-u)$$
$$B_{3,3} = (1-u)B_{3,2} + uB_{2,2} = u^3,$$

the third degree Bézier spline becomes

$$\boldsymbol{r}(u) = (1-u)^3\boldsymbol{V}_0 + 3u(1-u)^2\boldsymbol{V}_1 + 3u^2(1-u)\boldsymbol{V}_2 + u^3\boldsymbol{V}_3,$$

see Fig. 7.3.

For a Bézier spline the number of control vertices equals the degree of the curve plus one. Thus, in order to describe a complex boundary shape using only one or

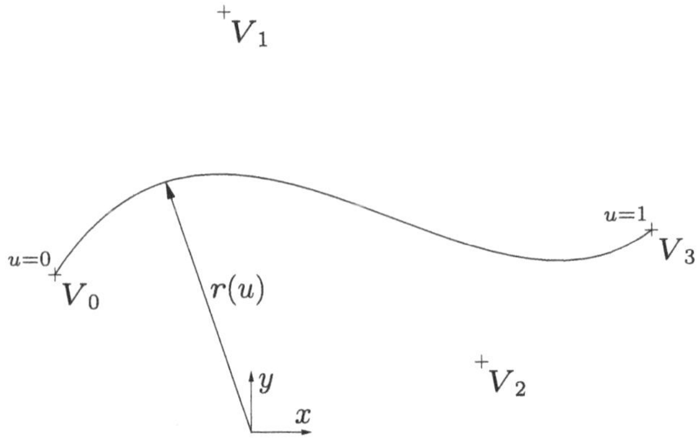

Fig. 7.3 Third degree Bézier spline

just a few Bézier splines, these curves have to be of high degree. Since this is impractical, one instead links together several low-degree splines imposing continuity constraints so that the boundary becomes sufficiently smooth. This will be illustrated in Sect. 7.2.

7.1.2 B-Splines

In contrast to a Bézier spline, the degree of a B-spline is not determined by the number of control vertices. A B-spline of degree p with $n+1$ control vertices is defined as

$$\boldsymbol{r}(u) = \sum_{i=0}^{n} M_{i,p}(u)\boldsymbol{V}_i, \quad 0 \le u \le 1, \tag{7.8}$$

where the pth degree B-spline basis functions $M_{i,p}$ are defined recursively as

$$M_{i,0}(u) = \begin{cases} 1 & \text{if } u_i \le u < u_{i+1} \\ 0 & \text{otherwise,} \end{cases} \tag{7.9}$$

$$M_{i,p}(u) = \frac{u - u_i}{u_{i+p} - u_i} M_{i,p-1}(u) + \frac{u_{i+p+1} - u}{u_{i+p+1} - u_{i+1}} M_{i+1,p-1}(u), \quad p \ge 1, \tag{7.10}$$

$$M_{m-p-1,p}(1) = 1. \tag{7.11}$$

The given scalars $u_0, u_1, \ldots, u_m$ are called knots. The number of knots, $m+1$, equals $p+n+2$. In order that the curve starts at $\boldsymbol{V}_0$ and ends at $\boldsymbol{V}_n$, the first $p+1$ knots are put to 0, and the last $p+1$ knots are put to 1. A knot vector $\mathcal{U}$ containing the knots is defined as

$$\mathcal{U} = \{\underbrace{0, \ldots, 0}_{p+1}, u_{p+1}, \ldots, u_{m-p-1}, \underbrace{1, \ldots, 1}_{p+1}\}.$$

In (7.10), $(u - u_i)/(u_{i+p} - u_i)$ and $(u_{i+p+1} - u)/(u_{i+p+1} - u_{i+1})$ should be interpreted as zero whenever $u_{i+p} - u_i = 0$ and $u_{i+p+1} - u_{i+1} = 0$, respectively.

The pth degree B-spline basis functions $M_{i,p}$ are only nonzero in a certain interval: $M_{i,p}(u) = 0$ for $u < u_i$ and $u \geq u_{i+p+1}$. Consequently, if the control vertex $\boldsymbol{V}_i$ is moved, the B-spline is only changed for $[u_i, u_{i+p+1})$; one says that B-splines have *local control*. Bézier splines, on the other hand, do not share this nice feature since a change in $\boldsymbol{V}_i$ will result in a change of the entire spline from start point to end point; Bézier curves have *global control*.

A B-spline of degree p is infinitely continuously differentiable at each u except at the knots. If knot j has multiplicity r_j, i.e. there are r_j knots with the same knot value u_j, then the curve is $p - r_j$ times continuously differentiable (C^{p-r_j}) at $u = u_j$. By altering the knot vector, low-order B-spline basis functions may be joined together *automatically* to form a complex curve with a desired degree of smoothness. This will be illustrated in several examples below. Recall that when low-order Bézier splines are used to represent a complex shape, the splines have to be joined together manually by imposing continuity constraints such that the curve is sufficiently smooth at the joints. For this reason B-splines are better suited for shape optimization than Bézier splines.

Just as for Bézier splines, the derivatives at the end points of a B-spline only depends on the location of the first and last two control vertices. It may be shown that

$$\frac{d\boldsymbol{r}(0)}{du} = \frac{p}{u_{p+1}}(\boldsymbol{V}_1 - \boldsymbol{V}_0) \tag{7.12}$$

$$\frac{d\boldsymbol{r}(1)}{du} = \frac{p}{1 - u_{m-p-1}}(\boldsymbol{V}_n - \boldsymbol{V}_{n-1}). \tag{7.13}$$

Example 7.1 We study a B-spline with three control vertices ($n = 2$):

$$\boldsymbol{V}_0 = \begin{bmatrix} 0 \\ 0 \end{bmatrix}, \quad \boldsymbol{V}_1 = \begin{bmatrix} 1 \\ 1 \end{bmatrix}, \quad \boldsymbol{V}_2 = \begin{bmatrix} 2 \\ 0 \end{bmatrix}.$$

At first the spline is taken to be a 1st degree curve ($p = 1$). The number of knots should then be $n + p + 2 = 5$ where the first two knots are 0 and the last two knots are 1. The knot vector is chosen as $\mathcal{U} = \{0, 0, 1/2, 1, 1\}$. The 0th and 1st degree basis functions are obtained from (7.9)–(7.11) as

$$M_{0,0} = 0$$

$$M_{1,0} = \begin{cases} 1 & \text{if } 0 \leq u < 1/2 \\ 0 & \text{otherwise} \end{cases}$$

$$M_{2,0} = \begin{cases} 1 & \text{if } 1/2 \le u < 1 \\ 0 & \text{otherwise} \end{cases}$$

$$M_{3,0} = 0$$

$$M_{0,1} = \frac{u-0}{0-0} M_{0,0} + \frac{1/2-u}{1/2-0} M_{1,0} = \begin{cases} 1-2u & \text{if } 0 \le u < 1/2 \\ 0 & \text{otherwise} \end{cases}$$

$$M_{1,1} = \frac{u-0}{1/2-0} M_{1,0} + \frac{1-u}{1-1/2} M_{2,0} = \begin{cases} 2u & \text{if } 0 \le u < 1/2 \\ 2(1-u) & \text{otherwise} \end{cases}$$

$$M_{2,1} = \frac{u-1/2}{1-1/2} M_{2,0} + \frac{1-u}{1-1} M_{3,0} = \begin{cases} 2u-1 & \text{if } 1/2 \le u \le 1 \\ 0 & \text{otherwise} \end{cases}$$

In Fig. 7.4, the 1st degree basis functions are plotted. Clearly, these functions are linearly independent, as suggested by their name. The B-spline $\boldsymbol{r}(\boldsymbol{u}) = \sum_{i=0}^{2} M_{i,1} \boldsymbol{V}_i$ is depicted in Fig. 7.5(a). Note that if (7.11) is not used, then (7.9) and (7.10) give $M_{2,1}(1) = 0$ which means that $\boldsymbol{r}(1) = \boldsymbol{0}$ rather than $\boldsymbol{r}(1) = \boldsymbol{V}_2$ as wanted. Looking at $M_{0,1}$ in Fig. 7.4, it is evident that a change in $\boldsymbol{V}_0$ will only affect the B-spline in the interval $[0,\ 1/2)$ which is in agreement with the statement above that a change in $\boldsymbol{V}_i$ only changes the curve in the interval $[u_i,\ u_{i+p+1})$. In Fig. 7.5(b), the B-spline is shown when $\boldsymbol{V}_0$ is changed to $\boldsymbol{V}_0 = [1\ 0]^T$.

Next, we change the B-spline into a 2nd degree curve ($p = 2$) by simply modifying the knot vector as $\mathcal{U} = \{0, 0, 0, 1, 1, 1\}$. The 0th, 1st and 2nd degree basis functions become

$$M_{0,0} = 0$$

$$M_{1,0} = 0$$

$$M_{2,0} = \begin{cases} 1 & \text{if } 0 \le u < 1 \\ 0 & \text{otherwise} \end{cases}$$

$$M_{3,0} = 0$$

$$M_{4,0} = 0$$

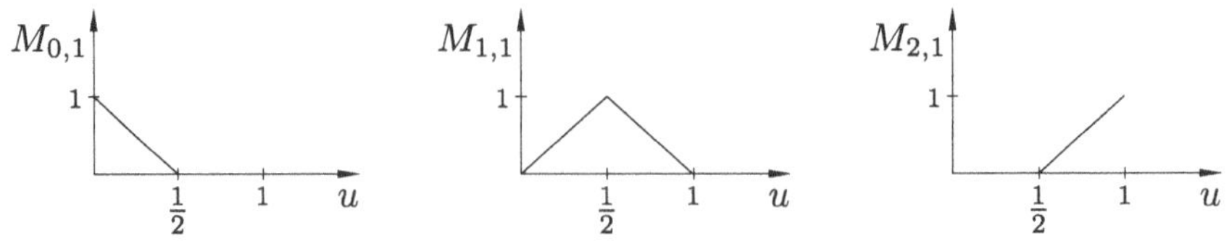

Fig. 7.4 The B-spline basis functions $M_{0,1}$, $M_{1,1}$ and $M_{2,1}$

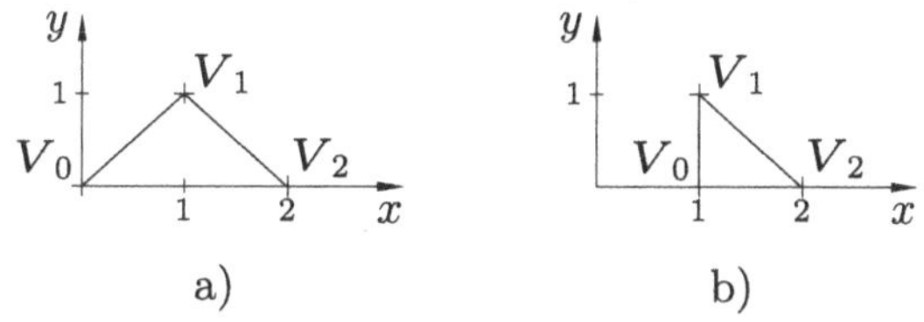

Fig. 7.5 A 1st degree B-spline

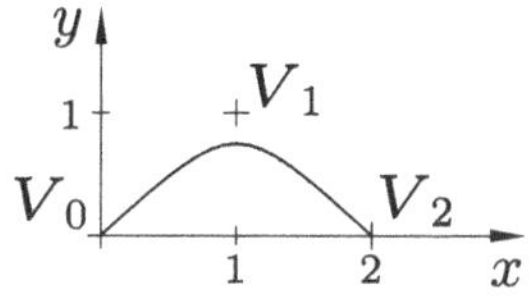

Fig. 7.6 A 2nd degree B-spline that is a Bézier spline

$$M_{0,1} = \frac{u-0}{0-0}M_{0,0} + \frac{0-u}{0-0}M_{1,0} = 0$$

$$M_{1,1} = \frac{u-0}{0-0}M_{1,0} + \frac{1-u}{1-0}M_{2,0} = \begin{cases} 1-u & \text{if } 0 \le u < 1 \\ 0 & \text{otherwise} \end{cases}$$

$$M_{2,1} = \frac{u-0}{1-0}M_{2,0} + \frac{1-u}{1-1}M_{3,0} = \begin{cases} u & \text{if } 0 \le u < 1 \\ 0 & \text{otherwise} \end{cases}$$

$$M_{3,1} = \frac{u-0}{1-1}M_{3,0} + \frac{1-u}{1-1}M_{4,0} = 0$$

$$M_{0,2} = \frac{u-0}{0-0}M_{0,1} + \frac{1-u}{1-0}M_{1,1} = \begin{cases} (1-u)^2 & \text{if } 0 \le u < 1 \\ 0 & \text{otherwise} \end{cases}$$

$$M_{1,2} = \frac{u-0}{1-0}M_{1,1} + \frac{1-u}{1-0}M_{2,1} = \begin{cases} 2u(1-u) & \text{if } 0 \le u < 1 \\ 0 & \text{otherwise} \end{cases}$$

$$M_{2,2} = \frac{u-0}{1-0}M_{2,1} + \frac{1-u}{1-1}M_{3,1} = \begin{cases} u^2 & \text{if } 0 \le u \le 1 \\ 0 & \text{otherwise.} \end{cases}$$

The B-spline then becomes

$$\boldsymbol{r}(\boldsymbol{u}) = \sum_{i=0}^{2} M_{i,2}\boldsymbol{V}_i = (1-u)^2\boldsymbol{V}_0 + 2u(1-u)\boldsymbol{V}_1 + u^2\boldsymbol{V}_2 = \begin{bmatrix} 2u \\ 2u(1-u) \end{bmatrix},$$

see Fig. 7.6. If we compare this with (7.7), we conclude that the curve is a 2nd degree Bézier spline! This is not a coincidence; any B-spline with $n = p$ and the knot vector

$$\mathcal{U} = \{\underbrace{0,\ldots,0}_{p+1},\underbrace{1,\ldots,1}_{p+1}\}.$$

is indeed a pth degree Bézier spline.

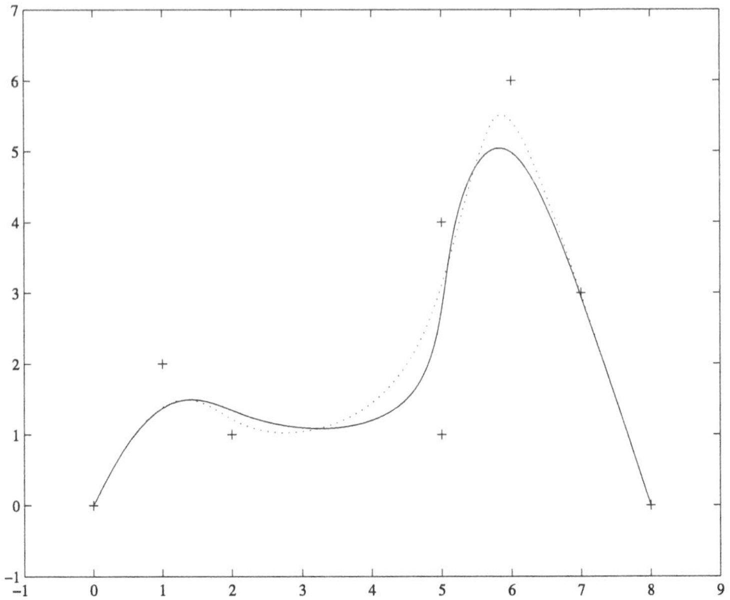

Fig. 7.7 Effect of knot spacing. *Solid line*: uniform knot vector. *Dotted line*: nonuniform knot vector

Example 7.2 A B-spline has the following eight control vertices ($n = 7$):

$$\boldsymbol{V}_0 = \begin{bmatrix} 0 \\ 0 \end{bmatrix}, \quad \boldsymbol{V}_1 = \begin{bmatrix} 1 \\ 2 \end{bmatrix}, \quad \boldsymbol{V}_2 = \begin{bmatrix} 2 \\ 1 \end{bmatrix}, \quad \boldsymbol{V}_3 = \begin{bmatrix} 5 \\ 1 \end{bmatrix},$$
$$\boldsymbol{V}_4 = \begin{bmatrix} 5 \\ 4 \end{bmatrix}, \quad \boldsymbol{V}_5 = \begin{bmatrix} 6 \\ 6 \end{bmatrix}, \quad \boldsymbol{V}_6 = \begin{bmatrix} 7 \\ 3 \end{bmatrix}, \quad \boldsymbol{V}_7 = \begin{bmatrix} 8 \\ 0 \end{bmatrix}.$$

For a 3rd degree spline ($p = 3$) the number of knots should be $n + p + 2 = 12$. Naturally, the knot sequence will influence the shape of the curve. If the knots in the interior of the curve, i.e. those strictly greater than 0 and strictly smaller than 1, are evenly spaced, the knot vector is said to be *uniform*. This does not imply that the distance between the points on the curve corresponding to the knots is the same since the arclength of a B-spline is not a linear function of u except for 1st degree curves. In Fig. 7.7 the splines corresponding to two different knot vectors—one uniform and the other nonuniform—are plotted. The knot vectors are

$$\{0,\ 0,\ 0,\ 0,\ 0.2,\ 0.4,\ 0.6,\ 0.8,\ 1,\ 1,\ 1,\ 1\}$$
$$\{0,\ 0,\ 0,\ 0,\ 0.1,\ 0.2,\ 0.8,\ 0.9,\ 1,\ 1,\ 1,\ 1\}.$$

In the upcoming shape optimization examples only uniform knot vectors will be used.

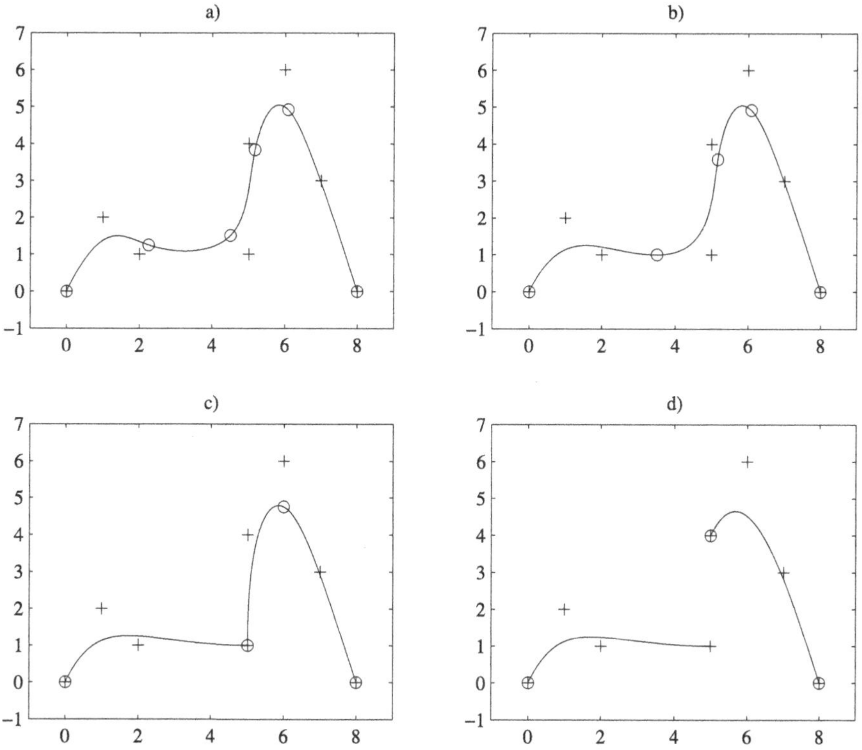

Fig. 7.8 Effect of multiple knots. Single knots in (**a**), a double knot in (**b**), a triple knot in (**c**) and a quadruple knot in (**d**). A *circle* is drawn at the points on the *curve* that correspond to a knot

In Fig. 7.8, the 3rd degree spline is drawn for knot vectors with multiple knots:

$$\mathcal{U}_a = \{0,\ 0,\ 0,\ 0,\ 1/5,\ 2/5,\ 3/5,\ 4/5,\ 1,\ 1,\ 1,\ 1\}$$
$$\mathcal{U}_b = \{0,\ 0,\ 0,\ 0,\ 1/4,\ 1/4,\ 2/4,\ 3/4,\ 1,\ 1,\ 1,\ 1\}$$
$$\mathcal{U}_c = \{0,\ 0,\ 0,\ 0,\ 1/3,\ 1/3,\ 1/3,\ 2/3,\ 1,\ 1,\ 1,\ 1\}$$
$$\mathcal{U}_d = \{0,\ 0,\ 0,\ 0,\ 1/2,\ 1/2,\ 1/2,\ 1/2,\ 1,\ 1,\ 1,\ 1\}.$$

The curve is C^{p-r} continuous at a joint with multiplicity r. Thus, for $\mathcal{U}_a$ the curve is C^2 at all knots, C^1 at $u = 1/4$ for $\mathcal{U}_b$, C^0 at $u = 1/3$ for $\mathcal{U}_c$, and discontinuous at $u = 1/2$ for $\mathcal{U}_d$.

For the 3rd degree spline with the uniform knot vector $\mathcal{U}_a$, control vertex $\boldsymbol{V}_1$ is moved from $[1\ 2]^T$ to $[1\ 3]^T$. This will change the curve in the interval $[u_1\ u_{1+p+1}) = [0\ 2/5)$, cf. Fig. 7.9.

The spline is changed into a 1st degree ($p = 1$) curve by altering the number of knots to $7 + 1 + 2 = 10$. A uniform knot vector is chosen, i.e.

$$\{0,\ 0,\ 1/7,\ 2/7,\ 3/7,\ 4/7,\ 5/7,\ 6/7,\ 1,\ 1\}.$$

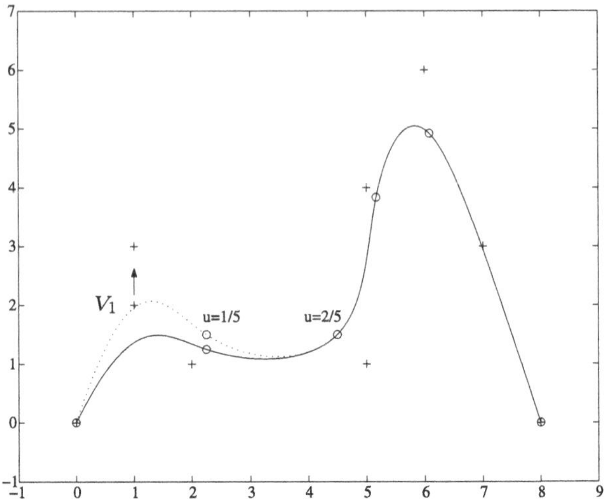

Fig. 7.9 Effect of changing the location of $\boldsymbol{V}_1$ for a 3rd degree B-spline. *Solid line*: $\boldsymbol{V}_1 = [1\ 2]^T$. *Dotted line*: $\boldsymbol{V}_1 = [1\ 3]^T$

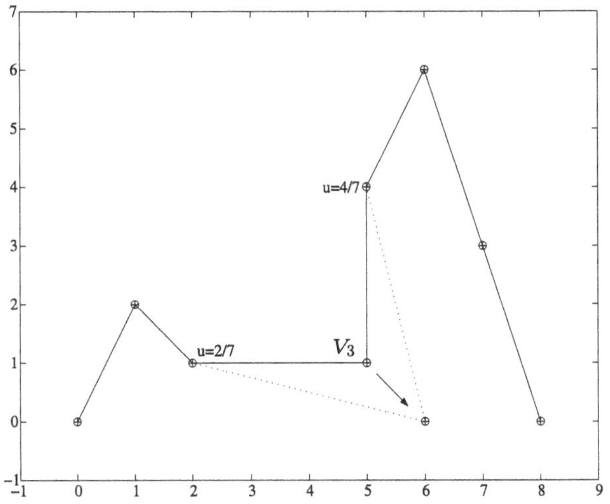

Fig. 7.10 Effect of changing the location of $\boldsymbol{V}_3$ for a 1st degree B-spline

The spline is drawn in Fig. 7.10 together with the spline obtained when the control vertex $\boldsymbol{V}_3 = [5\ 1]^T$ is moved to $[6\ 0]^T$. Note that a 1st degree B-spline is made up of lines connecting the control vertices, and that the knots will correspond to the control vertices. Moving $\boldsymbol{V}_i$ will only change the curve between $\boldsymbol{V}_{i-1}$ and $\boldsymbol{V}_{i+1}$.

Finally, the B-spline is changed to a 7th degree Bézier spline by using the knot vector

$$\{0,\ 0,\ 0,\ 0,\ 0,\ 0,\ 0,\ 0,\ 1,\ 1,\ 1,\ 1,\ 1,\ 1,\ 1,\ 1\}.$$

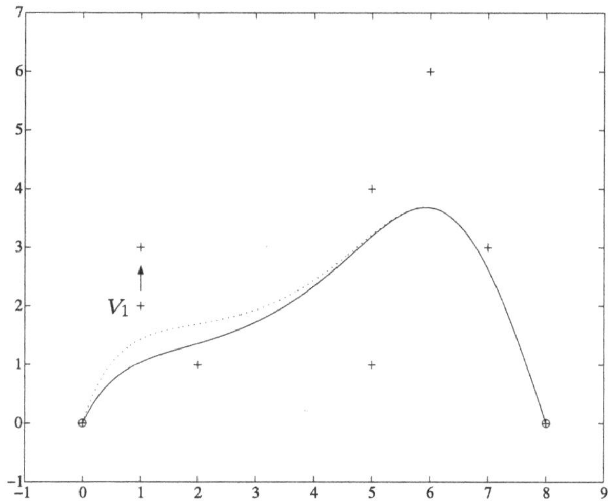

Fig. 7.11 A 7th degree Bézier spline modified by moving $\boldsymbol{V}_1$

In Fig. 7.11 this spline is shown together with the spline obtained when $\boldsymbol{V}_1$ is moved from $[1\ 2]^T$ to $[1\ 3]^T$. This movement results in the whole spline changing its shape although the spline is close to the original one far away from the vertex moved.

7.2 Treatment of Geometrical Design Constraints

When splines are linked together to form a boundary curve of a body, one has to impose constraints on some of the control vertices defining the splines such that the boundary has the desired degree of smoothness. Such constraints are almost inevitable if low-order Bézier splines are used, but could be necessary also when B-splines are used. Since the location of the control vertices is directly governed by design variables, the design variables will evidently be subject to constraints.

In Fig. 7.12 an axisymmetric body is shown. Given some objective function and constraints, one wants to find the optimum shape of the upper part of the body. For that purpose one half of the upper boundary is modeled using two 3rd degree Bézier splines. Spans and corresponding design variables $\alpha_1, \ldots, \alpha_8$ are also shown in the figure. The upper boundary should be C^1 continuous, which means that we have to impose constraints such that the boundary is C^1 continuous at the joint P_2 between the two splines, and that the boundary curve is C^1 continuous at P_1 for the complete structure, not just the half modeled. We will see shortly that there are two constraint equations at P_2 and one constraint equation at P_1 that may symbolically be written as

$$f_1(\boldsymbol{V}_1, \boldsymbol{V}_2) = 0$$

$$\boldsymbol{f}_2(\boldsymbol{V}_3, \boldsymbol{V}_4, \boldsymbol{V}_5) = \boldsymbol{0},$$

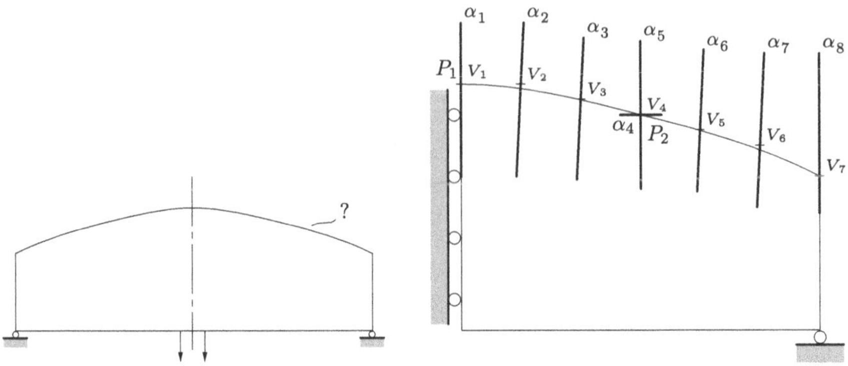

Fig. 7.12 An axisymmetric body to be optimized

where $\boldsymbol{V}_1 = \boldsymbol{V}_1(\alpha_1)$, $\boldsymbol{V}_2 = \boldsymbol{V}_2(\alpha_2)$, $\boldsymbol{V}_3 = \boldsymbol{V}_3(\alpha_3)$, $\boldsymbol{V}_4 = \boldsymbol{V}_4(\alpha_4, \alpha_5)$ and $\boldsymbol{V}_5 = \boldsymbol{V}_5(\alpha_6)$. One way of handling these equality constraints is to supply them to the optimization solver used. However, many implementations of optimization solvers for structural optimization are not able to handle equality constraints directly. Therefore, we choose to use the three equations to eliminate three design variables. Both f_1 and $\boldsymbol{f}_2$ turn out to be linear functions which makes it easy to write $\boldsymbol{V}_2 = \boldsymbol{V}_2(\alpha_1)$ and $\boldsymbol{V}_4 = \boldsymbol{V}_4(\alpha_3, \alpha_6)$, thus eliminating α_2, α_4 and α_5; the optimization solver will only use the independent design variables α_1, α_3, α_6, α_7 and α_8. The sensitivity of $\boldsymbol{V}_2(\alpha_1)$ and $\boldsymbol{V}_4(\alpha_3, \alpha_6)$ with respect to the independent design variables is obtained by differentiating these functions. For nonlinear constraint equations we cannot in general use the constraint equations to eliminate dependent design variables directly. Instead we obtain the needed sensitivities by differentiating the constraint equations themselves, as will be described in Sect. 7.2.3.

Next, we derive the constraint equations for the two continuity requirements discussed above, as well as for a composite circular arc where the constraint equations turn out to be nonlinear.

7.2.1 C^1 Continuity Between Bézier Splines

Two Bézier splines of the same degree are linked at the control vertex $\boldsymbol{V}_j$, see Fig. 7.13. The two vertices adjacent to $\boldsymbol{V}_j$ are denoted $\boldsymbol{V}_l$ and $\boldsymbol{V}_r$. The span corresponding to $\boldsymbol{V}_l$ is denoted $\boldsymbol{L}_l$ and the design variable that determines where $\boldsymbol{V}_l$ is located on the span $\boldsymbol{L}_l$ is denoted α_l. The span $\boldsymbol{L}_r$ and the design variable α_r correspond to vertex $\boldsymbol{V}_r$. From (7.5) and (7.6) we have that the composite curve is C^1 continuous at $\boldsymbol{V}_j$ if

$$\boldsymbol{V}_j - \boldsymbol{V}_l = \boldsymbol{V}_r - \boldsymbol{V}_j, \tag{7.14}$$

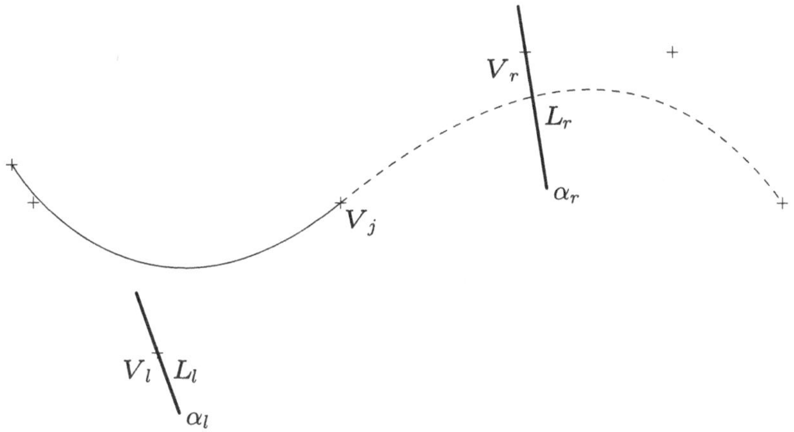

Fig. 7.13 Two linked Bézier splines

i.e.

$$V_j = \frac{1}{2}(V_l + V_r), \tag{7.15}$$

so that V_j is located right between V_l and V_r. For given values of α_l and α_r we may easily obtain the location of V_l and V_r using (7.1).

In order to calculate the nodal sensitivities, we first need to know the sensitivities of the control vertices. For control vertices independent of the location of other vertices, the sensitivity is given by (7.2), and for V_j the sensitivities may be obtained by differentiating (7.15) upon first using (7.2):

$$\frac{\partial V_l}{\partial \alpha_l} = L_l, \qquad \frac{\partial V_l}{\partial \alpha_r} = \mathbf{0}$$
$$\frac{\partial V_r}{\partial \alpha_l} = \mathbf{0}, \qquad \frac{\partial V_r}{\partial \alpha_r} = L_r.$$

Insertion of this into (7.15) yields the sensitivities

$$\frac{\partial V_j}{\partial \alpha_l} = \frac{1}{2} L_l$$
$$\frac{\partial V_j}{\partial \alpha_r} = \frac{1}{2} L_r. \tag{7.16}$$

All other $\partial V_j / \partial \alpha_k = \mathbf{0}$, $k \neq r,\ l$.

7.2.2 C^1 Continuity at a Point on a Line of Symmetry

The vertex V_j in Fig. 7.14 lies on a line of symmetry. The two adjacent vertices are V_l and V_r, where V_l is not actually modeled if we choose to take advantage of the

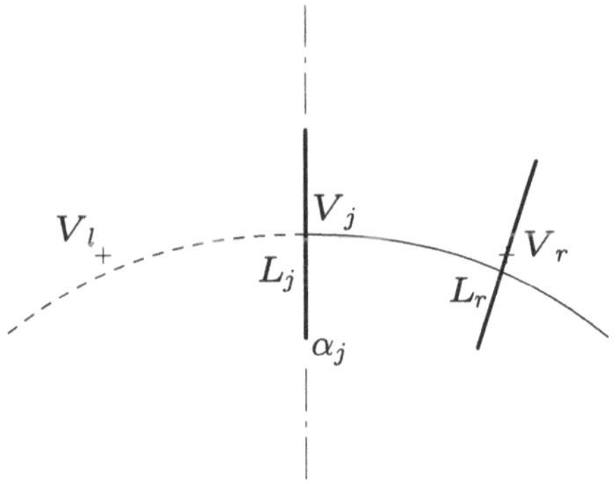

Fig. 7.14 C^1 continuity at vertex $\boldsymbol{V}_j$ located on the *symmetry line*

symmetry. If $\boldsymbol{V}_r = [x_r \; y_r]^T$, it thus holds that $\boldsymbol{V}_j = [0 \; y_j]^T$, $\boldsymbol{V}_l = [-x_r \; y_r]^T$. Since the two splines are of the same degree due to symmetry, the continuity constraints are the ones in (7.14). Note that (7.12) and (7.13) imply that the constraints are the same for any B-spline. By symmetry, (7.14) becomes

$$\begin{bmatrix} 0 \\ 2y_j - 2y_r \end{bmatrix} = \begin{bmatrix} 0 \\ 0 \end{bmatrix},$$

i.e. there is in fact only *one* constraint, namely that $\boldsymbol{V}_j$ and $\boldsymbol{V}_r$ should be located on the same y-coordinate. The location of $\boldsymbol{V}_j$ is given by

$$y_j = y_j^{\min} + \alpha_j L_{j,y},$$

and $\boldsymbol{V}_r$ is located on the line

$$x_r - x_r^{\min} = \frac{L_{r,x}}{L_{r,y}} \left(y_r - y_r^{\min} \right).$$

Since we must have that $y_r = y_j$, the location of $\boldsymbol{V}_r$ for any value of α_j is given by

$$y_r = y_j^{\min} + \alpha_j L_{j,y} \tag{7.17}$$

and

$$x_r = x_r^{\min} + \frac{L_{r,x}}{L_{r,y}} \left(y_j^{\min} + \alpha_j L_{j,y} - y_r^{\min} \right). \tag{7.18}$$

The sensitivity of the vertex $\boldsymbol{V}_r$ with respect to α_j then becomes

$$\frac{\partial x_r}{\partial \alpha_j} = L_{j,y} \frac{L_{r,x}}{L_{r,y}}$$

$$\frac{\partial y_r}{\partial \alpha_j} = L_{j,y},$$

i.e.

$$\frac{\partial \boldsymbol{V}_r}{\partial \alpha_j} = \frac{L_{j,y}}{L_{r,y}} \boldsymbol{L}_r, \tag{7.19}$$

and all other $\partial \boldsymbol{V}_r / \partial \alpha_k = \boldsymbol{0}$, $k \neq j$.

7.2.3 A Composite Circular Arc

Although very versatile, B-splines cannot be used to draw circles or any other conic sections. The more advanced NURBS can, however, be used to represent conic sections exactly. Since we do not cover NURBS in this work, we choose to treat circular arcs as special curves that may be used together with Bézier and B-splines to define boundary curves.

When using shorter circular arcs to draw a longer circular arc or a complete circle, one has to make sure that all end points of the smaller arcs lie on the same longer circular arc. To that end, study a circular arc with center $\boldsymbol{V}_c$, and end points $\boldsymbol{V}_s$ and $\boldsymbol{V}_j$, see Fig. 7.15. Another circular arc with center $\boldsymbol{V}_c$, and end points $\boldsymbol{V}_j$ and $\boldsymbol{V}_e$ is joined to the former arc at $\boldsymbol{V}_j$. A design variable α_s controls the location of $\boldsymbol{V}_s$ on the span $\boldsymbol{L}_s$. It is our aim to find out how $\boldsymbol{V}_j$ and $\boldsymbol{V}_e$ move along the spans $\boldsymbol{L}_j$ and $\boldsymbol{L}_e$, respectively, when α_s changes. For $\boldsymbol{V}_j$, the constraint equations are

$$|\boldsymbol{V}_s - \boldsymbol{V}_c| = |\boldsymbol{V}_j - \boldsymbol{V}_c|,$$

i.e.

$$(x_s - x_c)^2 + (y_s - y_c)^2 = (x_j - x_c)^2 + (y_j - y_c)^2, \tag{7.20}$$

and that $\boldsymbol{V}_j$ is located on the line

$$\begin{cases} y_j - y_j^{\min} = \dfrac{L_{j,y}}{L_{j,x}} \left(x_j - x_j^{\min} \right), & \text{if } L_{j,x} \neq 0 \\ x_j = x_j^{\min}, & \text{otherwise.} \end{cases} \tag{7.21}$$

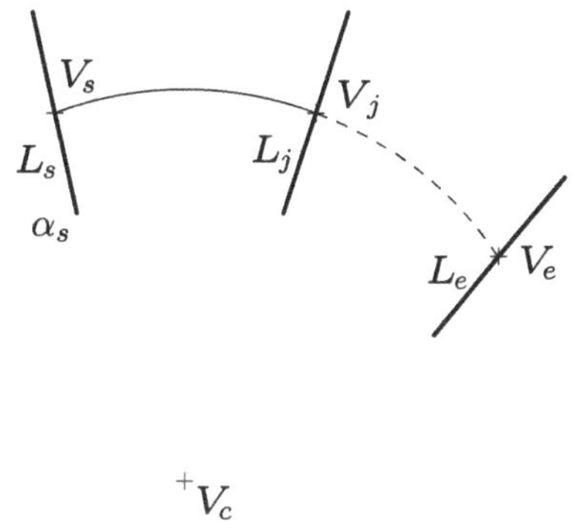

Fig. 7.15 Two circular arcs joined together

Since (7.20) is nonlinear, the easiest way to get the sensitivity of $\boldsymbol{V}_j$ is to differentiate the constraint equations rather than first trying to write $\boldsymbol{V}_j$ as a function of α_s and then differentiate. We get

$$(x_s - x_c)\frac{\partial x_s}{\partial \alpha_s} + (y_s - y_c)\frac{\partial y_s}{\partial \alpha_s} = (x_j - x_c)\frac{\partial x_j}{\partial \alpha_s} + (y_j - y_c)\frac{\partial y_j}{\partial \alpha_s},$$

and

$$\begin{cases} \dfrac{\partial y_j}{\partial \alpha_s} = \dfrac{L_{j,y}}{L_{j,x}}\dfrac{\partial x_j}{\partial \alpha_s}, & \text{if } L_{j,x} \neq 0 \\ \dfrac{\partial x_j}{\partial \alpha_s} = 0, & \text{otherwise,} \end{cases}$$

where $\partial x_s / \partial \alpha_s = L_{s,x}$ and $\partial y_s / \partial \alpha_s = L_{s,y}$. Solving these equations, we obtain the desired sensitivity as

$$\frac{\partial \boldsymbol{V}_j}{\partial \alpha_s} = \frac{(x_s - x_c)L_{s,x} + (y_s - y_c)L_{s,y}}{(x_j - x_c)L_{j,x} + (y_j - y_c)L_{j,y}}\boldsymbol{L}_j. \tag{7.22}$$

The denominator in this expression is nonzero as long as the span $\boldsymbol{L}_j$ is not tangent to the arc. Once the optimization solver has calculated a new value of α_s, $\boldsymbol{V}_s$ is obtained from (7.1). After that, (7.20) and (7.21) may be solved, either analytically or numerically, for $\boldsymbol{V}_j$. The constraint equations ensuring that $\boldsymbol{V}_e$ is located on the same arc as $\boldsymbol{V}_s$, and $\boldsymbol{V}_j$ are identical to those above, if the index j is replaced by e.

7.3 Mesh Generation and Calculation of Nodal Sensitivities

The nodal sensitivities will depend on how the finite element mesh is generated. There are two different classes of meshes: *structured* and *unstructured* meshes. In a structured, or *mapped*, mesh, there is a regular pattern of connections between the finite elements. Unstructured, or *free*, meshes are meshes that are not structured, see Fig. 7.16.

When creating a finite element model of a structure, one often first creates several regions bounded by some curves, and then meshes each region separately. Those regions whose shape will change during the optimization process are denoted *design elements*, cf. Fig. 7.17. Naturally, nodal sensitivities need to be calculated only for design elements since they are zero for other regions.

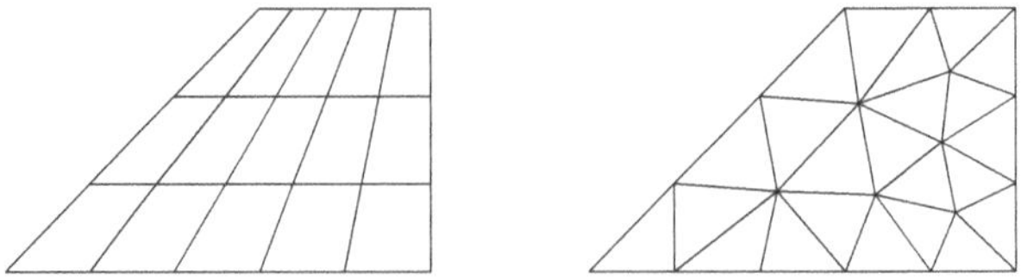

Fig. 7.16 A structured mesh to the *left*, and an unstructured mesh to the *right*

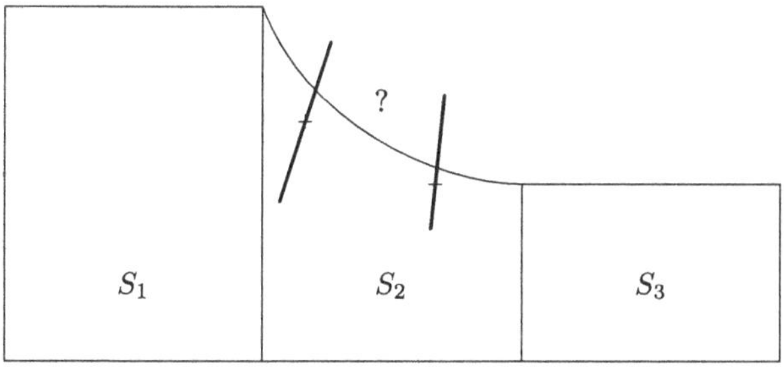

Fig. 7.17 A fillet to be optimized. The region S_2 is a design element

In what follows, we will describe how the nodal sensitivities may be obtained for two types of structured meshes, namely B-spline surface meshes and Coons surface meshes, as well as for unstructured meshes.

7.3.1 B-Spline Surface Meshes

A B-spline surface is defined as

$$\boldsymbol{r}(u, v) = \sum_{i=0}^{n} \sum_{j=0}^{m} M_{i,p}(u) M_{j,q}(v) \boldsymbol{V}_{i,j}, \quad 0 \le u \le 1, 0 \le v \le 1, \tag{7.23}$$

where the B-spline basis functions $M_{i,p}$ and $M_{j,q}$ using the knot vectors

$$\mathcal{U} = \{\underbrace{0, \ldots, 0}_{p+1}, u_{p+1}, \ldots, u_{r-p-1}, \underbrace{1, \ldots, 1}_{p+1}\}$$

$$\mathcal{V} = \{\underbrace{0, \ldots, 0}_{q+1}, v_{q+1}, \ldots, v_{s-q-1}, \underbrace{1, \ldots, 1}_{q+1}\},$$

are defined in (7.9)–(7.11). For a fixed $v = \bar{v}$, $\boldsymbol{r}(u, \bar{v})$ is a pth degree B-spline, and for a fixed $u = \bar{u}$, $\boldsymbol{r}(\bar{u}, v)$ is a qth degree B-spline. The number of knots in $\mathcal{U}$ and $\mathcal{V}$ is $r + 1$ and $s + 1$, respectively. It holds that $r = n + p + 1$ and $s = m + q + 1$. The vectors $\boldsymbol{V}_{i,j} = [x_{i,j}\ y_{i,j}]^T$, $i = 0, \ldots, n$, $j = 0, \ldots, m$, are control vertices. As before, there may be one or two spans defined for a control vertex so that it can move during the optimization process. If $\boldsymbol{V}_{i,j}$ is moved, the surface is changed only in the interval $[u_i,\ u_{i+p+1}) \times [v_j,\ v_{j+q+1})$. Specifically, if interior control vertices, i.e. vertices $\boldsymbol{V}_{i,j}$ with $1 \le i \le n-1$ and $1 \le j \le m-1$, are moved, the shape of the boundary curves is not affected.

It is possible to add a third coordinate to the control vertices which can be used to describe a nonconstant thickness of a sheet: $\boldsymbol{V}_{i,j} = [x_{i,j}\ y_{i,j}\ z_{i,j}]^T$. Unless all $z_{i,j}$ are the same, $\boldsymbol{r}(u, v)$ will be a curved surface.

Finite element nodes are created by evaluating $\boldsymbol{r}(u, v)$ for the u-values $u_0^n, \ldots, u_{n_u}^n$, where $0 = u_0^n < u_1^n < \cdots < u_{n_u-1}^n < u_{n_u}^n = 1$, and the v-values $v_0^n, \ldots, v_{n_v}^n$, where $0 = v_0^n < v_1^n < \cdots < v_{n_v-1}^n < v_{n_v}^n = 1$. The nodal sensitivities are obtained by differentiating the mapping (7.23) with respect to the independent design variables α_k, $k = 0, 1, \ldots,$ at the u- and v-values corresponding to the finite element nodes:

$$\frac{\partial \boldsymbol{r}(u, v)}{\partial \alpha_k} = \sum_{i=0}^{n} \sum_{j=0}^{m} M_{i,p}(u) M_{j,q}(v) \frac{\partial \boldsymbol{V}_{i,j}}{\partial \alpha_k}, \tag{7.24}$$

where $\partial \boldsymbol{V}_{i,j}/\partial \alpha_k$ is given by (7.2) for vertices independent of the location of other vertices. For vertices that are constrained by other vertices, $\partial \boldsymbol{V}_{i,j}/\partial \alpha_k$ is obtained following the procedure of Sect. 7.2, cf. e.g. (7.16), (7.19) or (7.22).

A disadvantage with any structured mesh is that even though the mesh is of good quality before the shape optimization starts, the elements may become more and more distorted as the optimization proceeds. If this happens, it is recommended to abort the optimization, remesh the structure, and continue the optimization with this new, improved mesh.

Example 7.3 A B-spline surface has $n = 3$, $p = 2$, $m = 2$, $q = 1$, $\mathcal{U} = \{0, 0, 0, 1/2, 1, 1, 1\}$, $\mathcal{V} = \{0, 0, 1/2, 1, 1\}$, and

$$\boldsymbol{V}_{0,0} = \begin{bmatrix} 0 \\ 0 \end{bmatrix}, \quad \boldsymbol{V}_{1,0} = \begin{bmatrix} 2 \\ 0 \end{bmatrix}, \quad \boldsymbol{V}_{2,0} = \begin{bmatrix} 4 \\ 0 \end{bmatrix}, \quad \boldsymbol{V}_{3,0} = \begin{bmatrix} 6 \\ 0 \end{bmatrix}$$

$$\boldsymbol{V}_{0,1} = \begin{bmatrix} -0.2 \\ 1.6 \end{bmatrix}, \quad \boldsymbol{V}_{1,1} = \begin{bmatrix} 2 \\ 2 \end{bmatrix}, \quad \boldsymbol{V}_{2,1} = \begin{bmatrix} 4 \\ 2 \end{bmatrix}, \quad \boldsymbol{V}_{3,1} = \begin{bmatrix} 6.5 \\ 2 \end{bmatrix}$$

$$\boldsymbol{V}_{0,2} = \begin{bmatrix} 0 \\ 3.2 \end{bmatrix}, \quad \boldsymbol{V}_{1,2} = \begin{bmatrix} 2 \\ 2.8 \end{bmatrix}, \quad \boldsymbol{V}_{2,2} = \begin{bmatrix} 5 \\ 4.4 \end{bmatrix}, \quad \boldsymbol{V}_{3,2} = \begin{bmatrix} 6 \\ 4 \end{bmatrix}.$$

A mesh is created by evaluating $\boldsymbol{r}(u, v)$ for $u = 0, 0.025, 0.05, \ldots, 1$ and $v = 0, 0.05, 0.1, \ldots, 1$. Each curve in the mesh corresponds to a constant value of u or v, see Fig. 7.18. The control vertex $\boldsymbol{V}_{0,2}$ is then moved from $[0\ 3.2]^T$ to $[0\ 4.2]^T$. This will change the B-spline surface only for $[u_i\ u_{i+p+1}) \times [v_j\ v_{j+q+1}) = [0\ 1/2) \times [1/2\ 1)$, as illustrated in Fig. 7.19.

7.3.2 *Coons Surface Meshes*

In a B-spline surface, opposite boundary curves are necessarily of the same degree, and controlled by the same number of control vertices. It would be nice to be able to choose all four boundary curves arbitrarily. Coons surfaces provide this possibility. A four-sided[1] Coons surface is defined as

$$\boldsymbol{r}(u, v) = (1 - u)\boldsymbol{\rho}_0(v) + u\boldsymbol{\rho}_1(v) + (1 - v)\boldsymbol{r}_0(u) + v\boldsymbol{r}_1(u)$$

[1] It is also possible to define three-sided Coons surfaces.

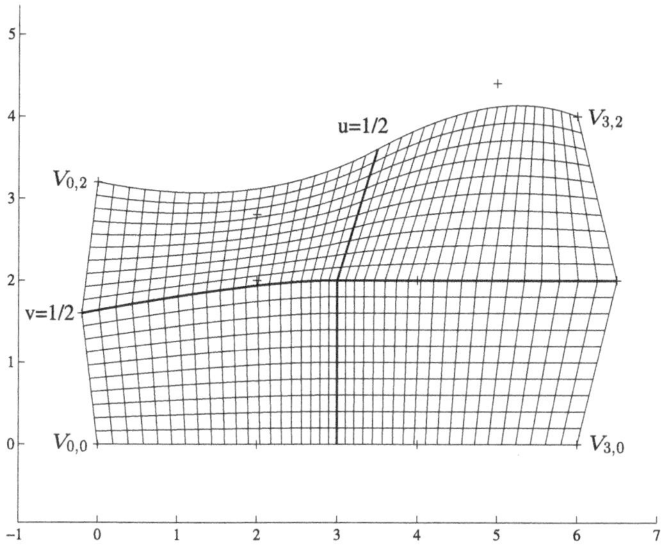

Fig. 7.18 A B-spline surface mesh

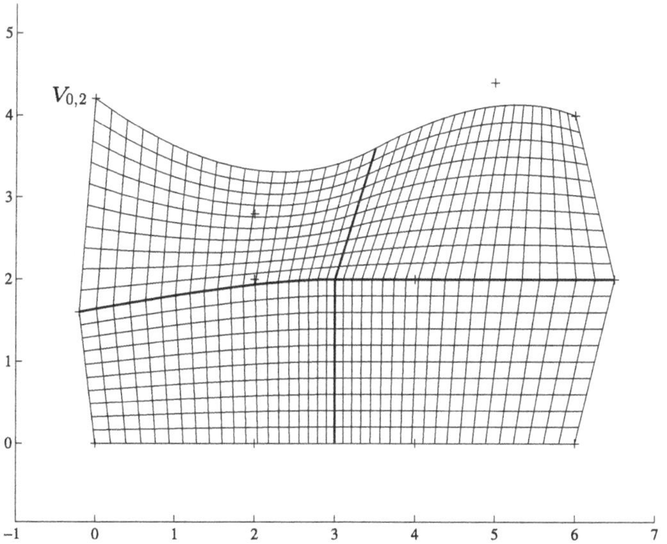

Fig. 7.19 Effect of moving $\boldsymbol{V}_{0,2}$ from $[0\ 3.2]^T$ to $[0\ 4.2]^T$

$$- (1-u)(1-v)\boldsymbol{r}_{0,0} - (1-u)v\boldsymbol{r}_{0,1} - u(1-v)\boldsymbol{r}_{1,0} - uv\boldsymbol{r}_{1,1}, \tag{7.25}$$

where $\boldsymbol{r}_0(u)$, $\boldsymbol{r}_1(u)$, $\boldsymbol{\rho}_0(v)$ and $\boldsymbol{\rho}_1(v)$, $0 \le u \le 1$, $0 \le v \le 1$ are arbitrary boundary curves, and $\boldsymbol{r}_{0,0} = \boldsymbol{r}_0(0) = \boldsymbol{\rho}_0(0)$, $\boldsymbol{r}_{1,0} = \boldsymbol{r}_0(1) = \boldsymbol{\rho}_1(0)$, $\boldsymbol{r}_{0,1} = \boldsymbol{r}_1(0) = \boldsymbol{\rho}_0(1)$,

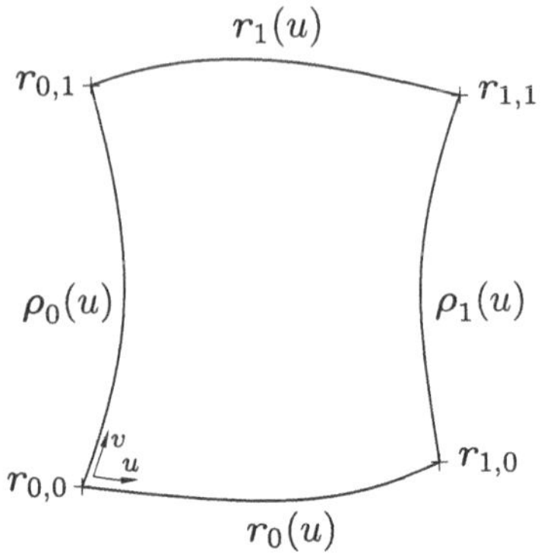

Fig. 7.20 Boundary curves for a Coons surface

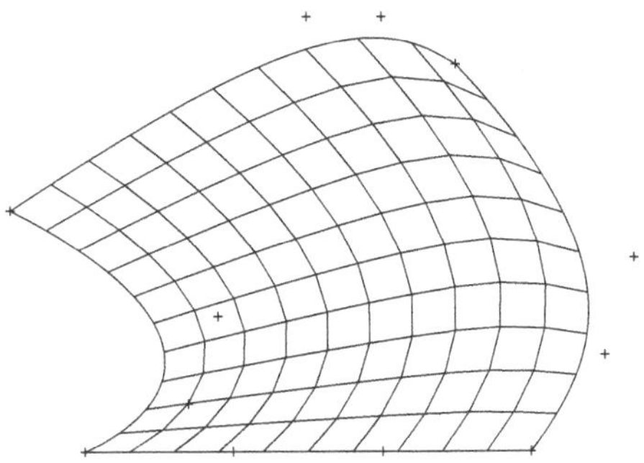

Fig. 7.21 Example of a Coons surface mesh

$\boldsymbol{r}_{1,1} = \boldsymbol{r}_1(1) = \boldsymbol{\rho}_1(1)$, see Fig. 7.20. Finite element nodes are obtained by evaluating $\boldsymbol{r}(u, v)$ for suitable u- and v-values in the same way as was done for B-spline surfaces. In Fig. 7.21, a Coons surface mesh is shown where the boundary curves are all 3rd degree Bézier splines. Nodal sensitivities are obtained by differentiating (7.25). Indicating $\partial/\partial\alpha_k$ by $'$ we get

$$\begin{aligned}\boldsymbol{r}'(u, v) = &(1-u)\boldsymbol{\rho}_0'(v) + u\boldsymbol{\rho}_1'(v) + (1-v)\boldsymbol{r}_0'(u) + v\boldsymbol{r}_1'(u)\\ &- (1-u)(1-v)\boldsymbol{r}_{0,0}' - (1-u)v\boldsymbol{r}_{0,1}' - u(1-v)\boldsymbol{r}_{1,0}' - uv\boldsymbol{r}_{1,1}',\end{aligned} \tag{7.26}$$

where, if e.g. $\boldsymbol{r}_0$ is a B-spline of degree p with $n+1$ control vertices, $\boldsymbol{r}_0'(u)$ is obtained from (7.8) as

$$\frac{\partial \boldsymbol{r}_0(u)}{\partial \alpha_k} = \sum_{i=0}^{n} M_{i,p}(u) \frac{\partial \boldsymbol{V}_i}{\partial \alpha_k}. \tag{7.27}$$

7.3.3 Unstructured Meshes

In order to be able to use structured meshes for bodies with a complex geometry, the body usually has to be divided into quite a number of four- or three-sided regions, each of which is then meshed. Thus, it can be very tedious to use a structured mesh generator for complex geometries. As discussed above, the mesh may also become severely distorted as the control vertices move during the optimization process. By using unstructured meshes instead, both of these problems may be avoided. The price to pay is a more complicated computer implementation.

There is a multitude of algorithms for generating unstructured meshes, which are all based on either so-called Delaunay, quadtree–octree, or advancing front techniques. This is not the place to present a review of all these techniques; interested readers are referred to Thompson, Soni and Weatherill [35], George [16] and Cheung, Lo and Leung [23] for details. For the sake of illustration, we will, however, describe *one* algorithm, namely the very simple advancing front algorithm by Tracy [36].

0. Nodes are placed along the boundary curves of the body. The nodes are interconnected by straight lines that together form the *front*, see Fig. 7.22(a).
1. For each corner angle $\angle ABC$ of the front that is less than 90°, an element ABC is created, where A and C are the nodes adjacent to B. The front is updated by replacing the line segment A–B–C with A–C, cf. Fig. 7.22(b).
2. At any corner angle $\angle ABC$ of the front less than 180°, where the line segment A–B–C runs anticlockwise, create a new node D according to the ad hoc rule

$$x_D = \frac{1}{2}(x_A + x_C) + \frac{1}{5}(y_A - y_C)$$

$$y_D = \frac{1}{2}(y_A + y_C) + \frac{1}{5}(x_C - x_A).$$

   Create the elements ABD and BCD. The front is updated by replacing the line segment A–B–C with A–D–C, as illustrated in Fig. 7.22(c).
3. Go to step 1. Stop when the whole body has been triangularized.

The algorithm is illustrated in Fig. 7.23. The quality of the mesh obtained may be improved by various techniques, such as the widely used Laplacian smoothing method. Here, the internal nodes in the mesh are relocated in order to make the

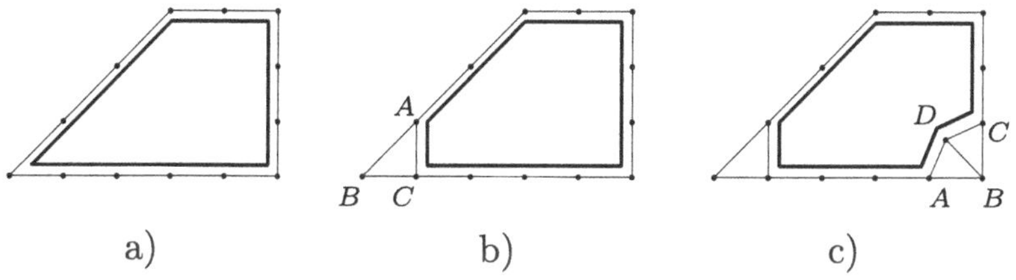

Fig. 7.22 The initial front with updates

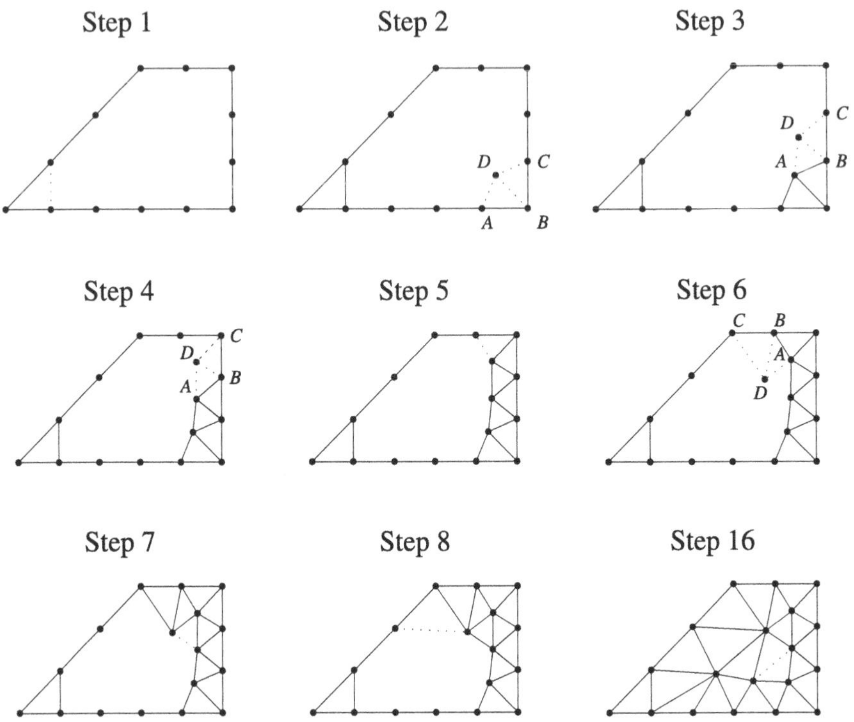

Fig. 7.23 A mesh created by Tracy's advancing front algorithm

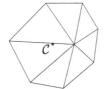

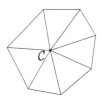

Fig. 7.24 Relocation of interior node with Laplacian smoothing

elements more equilateral. This is accomplished by moving each internal node to the centroid of the polygon comprised of the elements containing that node, see Fig. 7.24. One usually has to process the internal nodes of the mesh at least twice, i.e. perform two smoothing iterations, to see any significant improvement in mesh quality, cf. Fig. 7.25.

For a structured mesh, nodal sensitivities are obtained by simply differentiating the mesh function. For unstructured meshes there does not exist any continuous mesh function to differentiate since the mesh will change its topology as the body changes its shape. For nodes on the boundary curves, the nodal sensitivities may, as for structured meshes, be obtained by differentiating the functions defining the curves, cf. e.g. (7.27) for B-splines. For interior nodes things are more complicated. We refer to Bugeda and Oliver [11] for a discussion. Numerical examples have shown, however, see e.g. Hilding, Torstenfelt and Klarbring [20], that the sensi-

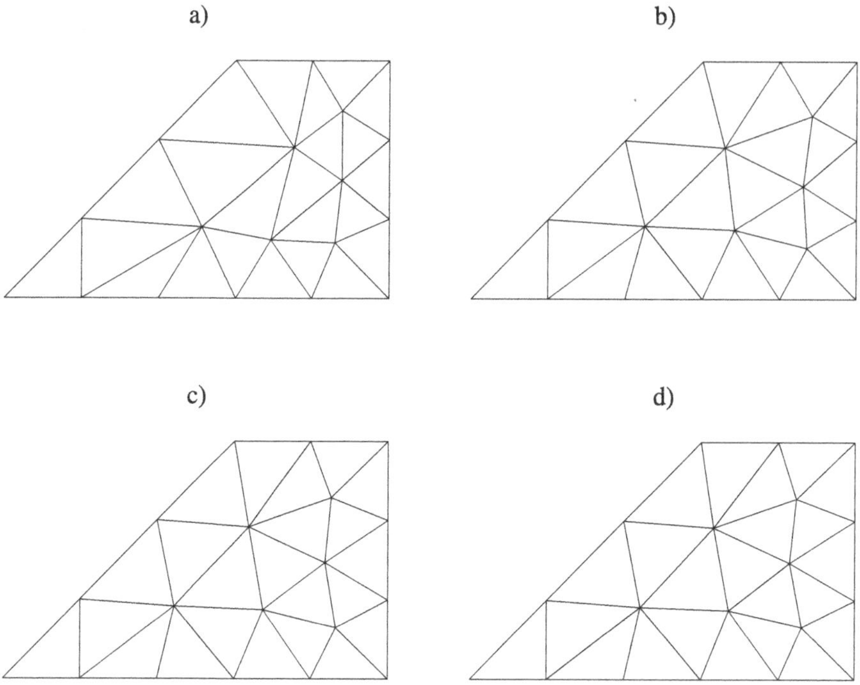

Fig. 7.25 Laplacian smoothing of a mesh; (**a**) no smoothing, (**b**) one smoothing iteration, (**c**) two iterations, (**d**) three iterations

tivities of interior nodes may be put to zero and still provide sufficiently accurate sensitivities of the objective function and constraints for the optimization solver! The reason that this works is probably that the mechanical behavior of a discretized structure is mostly dependent on the shape of its boundary curves, and to a lesser extent on the location of the interior nodes, at least for good quality meshes.

7.4 Summary of Sensitivity Analysis for Two-Dimensional Shape Optimization

For convenience, we summarize all steps needed to perform the sensitivity analysis in shape optimization of plane sheets using the direct analytical method. The nested optimization problem may be written

$$(\mathbb{SO})_{\mathrm{nf}} \quad \begin{cases} \min\limits_{\boldsymbol{\alpha}} & \hat{g}_0(\boldsymbol{\alpha}) = g_0(\boldsymbol{\alpha}, \boldsymbol{u}(\boldsymbol{\alpha})) \\ \text{s.t.} & \hat{g}_i(\boldsymbol{\alpha}) = g_i(\boldsymbol{\alpha}, \boldsymbol{u}(\boldsymbol{\alpha})) \leq 0, \quad i = 1, \ldots, l \\ & 0 \leq \alpha_j \leq 1, \quad j = 1, \ldots, n. \end{cases}$$

The sensitivities of $\hat{g}_i$, $i = 0, \ldots, l$, may be obtained as follows.

For each (independent) design variable α_j:

For each design element (DE) whose shape is affected by α_j:

Obtain the nodal sensitivities $\partial \boldsymbol{X}/\partial \alpha_j$ for all nodes in the DE.

For each finite element e in the DE:

Evaluate the sensitivity of the shape function derivatives matrix $\boldsymbol{G}$ and the Jacobian $\boldsymbol{J}$ at the Gauss points of the finite element by using (6.30) and (6.31):

$$\frac{\partial \boldsymbol{G}}{\partial \alpha_j} = -\boldsymbol{G}\frac{\partial \boldsymbol{X}}{\partial \alpha_j}\boldsymbol{G}$$

$$\frac{\partial |\boldsymbol{J}|}{\partial \alpha_j} = |\boldsymbol{J}|\,\mathrm{tr}\left(\boldsymbol{G}\frac{\partial \boldsymbol{X}}{\partial \alpha_j}\right).$$

Obtain the sensitivity of the strain displacement matrix $\boldsymbol{B}$, $\partial \boldsymbol{B}/\partial \alpha_j$, from $\partial \boldsymbol{G}/\partial \alpha_j$, cf. (6.22) and (6.20). Get the sensitivity of the element stiffness matrix and the element applied force vector from (6.21) and (6.33):

$$\frac{\partial \boldsymbol{k}_e}{\partial \alpha_j} = \int_{\hat{\Omega}} \left(\frac{\partial \boldsymbol{B}^T}{\partial \alpha_j}\boldsymbol{D}\boldsymbol{B}|\boldsymbol{J}| + \boldsymbol{B}^T\boldsymbol{D}\frac{\partial \boldsymbol{B}}{\partial \alpha_j}|\boldsymbol{J}| + \boldsymbol{B}^T\boldsymbol{D}\boldsymbol{B}\frac{\partial |\boldsymbol{J}|}{\partial \alpha_j}\right) t\, d\hat{\Omega}$$

$$\frac{\partial \boldsymbol{f}_e^a}{\partial \alpha_j} = \int_{\hat{\Omega}} \boldsymbol{N}^T \left(\frac{\partial \boldsymbol{b}}{\partial \alpha_j}|\boldsymbol{J}| + \boldsymbol{b}\frac{\partial |\boldsymbol{J}|}{\partial \alpha_j}\right) t\, d\hat{\Omega}.$$

Form the pseudo-load $\partial \boldsymbol{f}_e^a/\partial \alpha_j - (\partial \boldsymbol{k}_e/\partial \alpha_j)\boldsymbol{u}_e$ and assemble to form $\partial \boldsymbol{F}/\partial \alpha_j - (\partial \boldsymbol{K}/\partial \alpha_j)\boldsymbol{u}$, cf. (6.9).

end

end

Solve (6.4) for $\partial \boldsymbol{u}/\partial \alpha_j$:

$$\boldsymbol{K}\frac{\partial \boldsymbol{u}}{\partial \alpha_j} = \frac{\partial \boldsymbol{F}}{\partial \alpha_j} - \frac{\partial \boldsymbol{K}}{\partial \alpha_j}\boldsymbol{u}.$$

Use (6.3) to calculate the required sensitivities of the objective function and the constraints:

$$\frac{\partial \hat{g}_i}{\partial \alpha_j} = \frac{\partial g_i}{\partial \alpha_j} + \frac{\partial g_i}{\partial \boldsymbol{u}}\frac{\partial \boldsymbol{u}}{\partial \alpha_j},$$

for $i = 0, \ldots, l$.

end

Example 7.4 The shape of the upper boundary of the sheet in Fig. 7.26 should be determined so that the sheet's stiffness is maximized by minimizing its compliance. The weight of the sheet is not allowed to exceed 180 units, and the density is 1. The sheet is modeled as a B-spline surface with 6×2 control vertices,

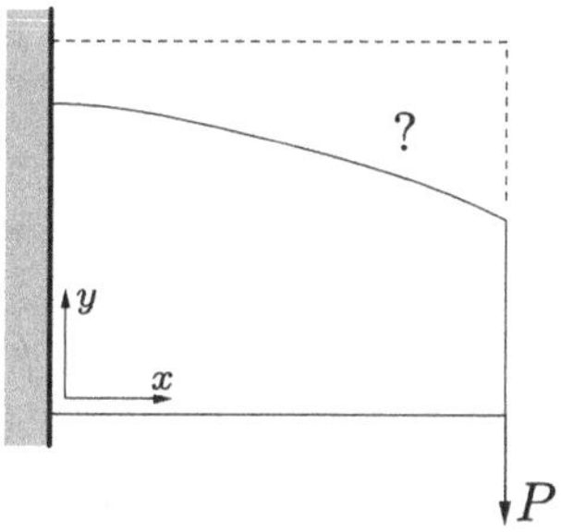

Fig. 7.26 The sheet to be optimized

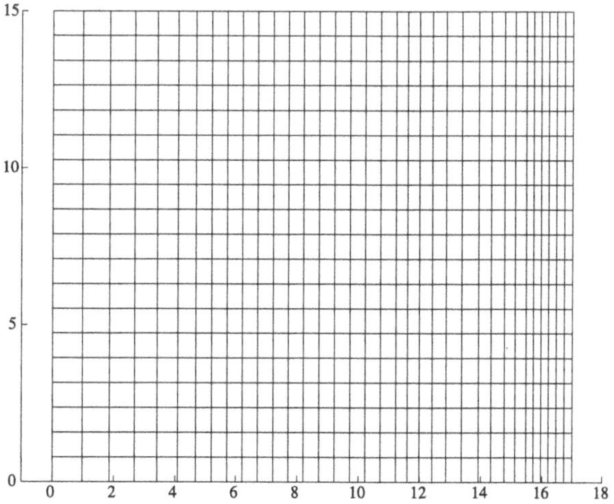

Fig. 7.27 Initial mesh of the sheet

2nd degree curves in the u-direction, and 1st degree curves in the v-direction. Initially, the control vertices defining the shape of the upper boundary are located at $x = 0,\ 3.4,\ 6.8,\ 10.2,\ 13.6,\ 17$ and $y = 15$. These vertices may move vertically between $y = 1.5$ and $y = 20$ during the optimization. Figure 7.27 shows the initial mesh. The optimization problem is solved using MMA. Figure 7.28 illustrates how the shape of the sheet changes during the optimization process. Already after 4–5 iterations, the shape hardly changes at all, see also Fig. 7.29 where the variation of the compliance and the weight during the optimization are shown. Note that the initial configuration is not feasible since the initial weight is greater than 180 units.

Example 7.5 The compliance of the fillet in Fig. 7.30 should be minimized. The initial weight is 0.32 units, but the optimized fillet should have a weight not greater than 0.25 units. Two interconnected cubic Bézier splines are allowed to change their shape. At the point where the two curves meet, the boundary is constrained to be G^1 continuous, which is a relaxation of the constraint of being C^1 continuous, see Exercise 7.2. The finite element program TRINITAS has been used to solve the problem. An unstructured mesh is created with 298 nodes and 534 elements. The unstructured mesh generator implemented in TRINITAS is a slight modification of Tracy's algorithm described in Sect. 7.3.3. TRINITAS uses MMA to solve the

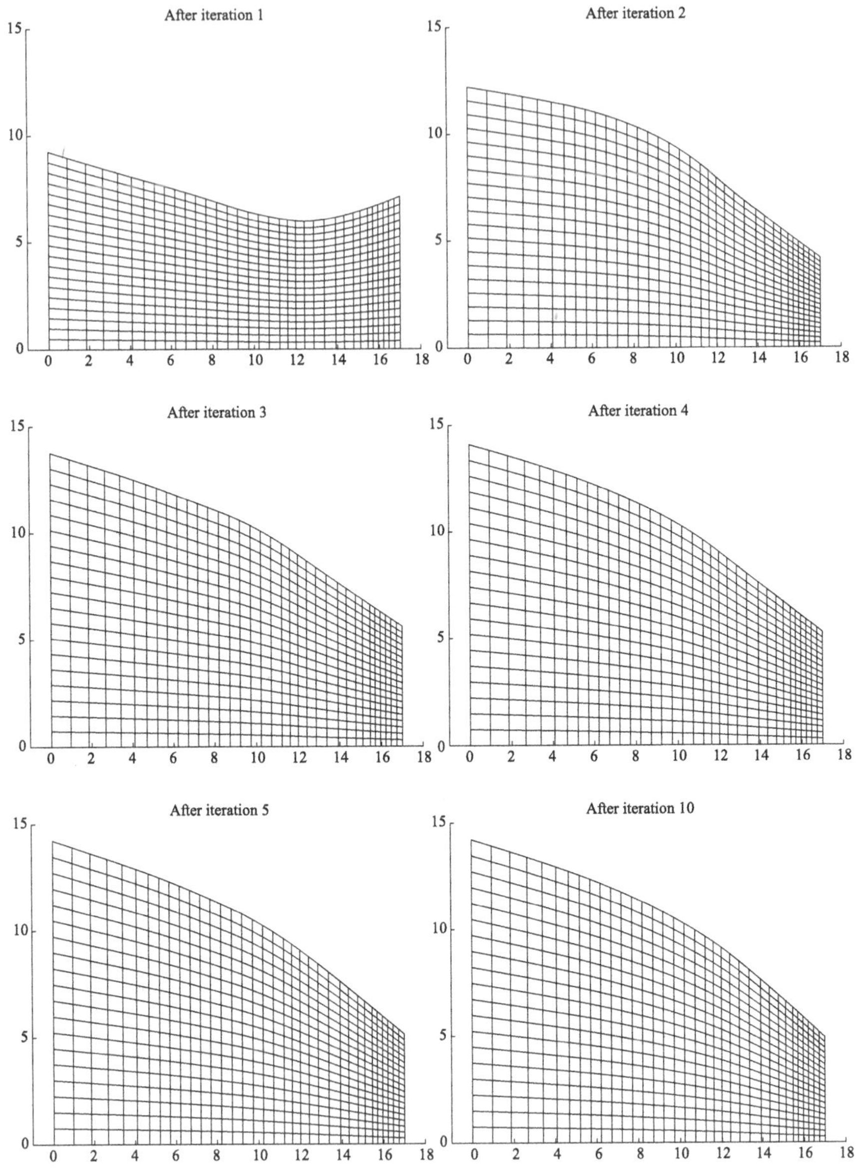

Fig. 7.28 The shape of the sheet at various iterations

optimization problem. The shape converges after some 5 iterations, see Fig. 7.31. Note that there are fewer nodes and elements in the final configuration than in the initial one: 228 nodes and 394 elements. On the boundary curves, however, the number of nodes does not change. Since the optimized structure is smaller than the initial, it is therefore obvious that the number of nodes must decrease when Tracy's algorithm is used as a mesh generator.

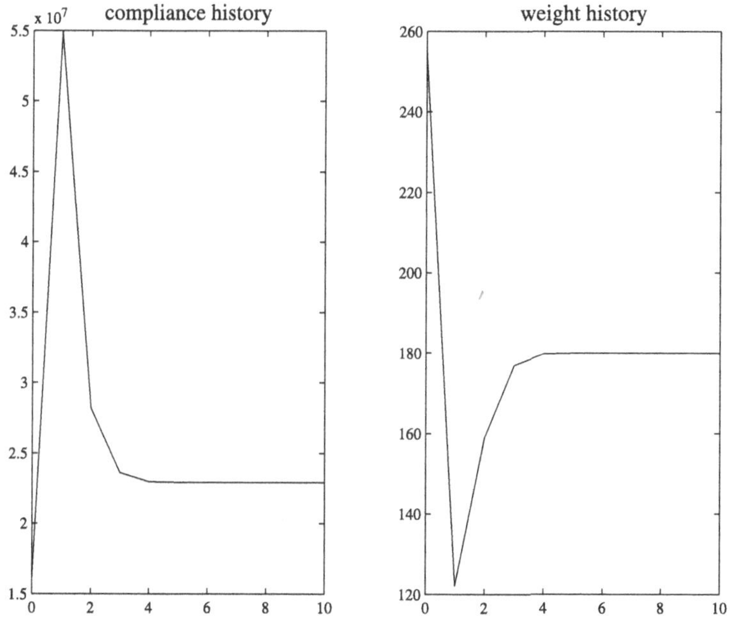

Fig. 7.29 The history of the compliance and the weight

7.5 Exercises

Exercise 7.1 Two B-splines are interconnected at $\boldsymbol{V}_j$, see Fig. 7.32. The two adjacent control vertices are $\boldsymbol{V}_l$ and $\boldsymbol{V}_r$. The spline which is controlled by $\boldsymbol{V}_l$ has degree p^l, and its greatest knot which is smaller than 1 is u^l. The other spline has degree p^r and its smallest knot greater than 0 is u^r. The composite curve should be C^1 continuous at $\boldsymbol{V}_j$. Show that the sensitivities of $\boldsymbol{V}_j$ are given by

$$\frac{\partial \boldsymbol{V}_j}{\partial \alpha_r} = \frac{(1-u^l)p^r}{(1-u^l)p^r + u^r p^l}\boldsymbol{L}_r$$

$$\frac{\partial \boldsymbol{V}_j}{\partial \alpha_l} = \frac{u^r p^l}{(1-u^l)p^r + u^r p^l}\boldsymbol{L}_l.$$

Exercise 7.2 Two Bézier splines of equal degree are interconnected at $\boldsymbol{V}_j$ as shown in Fig. 7.33. The composite curve is said to be G^1 *continuous* at $\boldsymbol{V}_j$ if the tangent $d\boldsymbol{r}/du$ of the curves point in the same direction at $\boldsymbol{V}_j$, but are not necessarily equal in magnitude (that is, the curve is not necessarily C^1 continuous at $\boldsymbol{V}_j$). Show that the curve is G^1 at $\boldsymbol{V}_j$ if, and only if, $\boldsymbol{V}_j$ lies on the line between $\boldsymbol{V}_l$ and $\boldsymbol{V}_r$. Recall that the curve is C^1 if, and only if, $\boldsymbol{V}_j$ lies at the mid-point of the line between $\boldsymbol{V}_l$ and $\boldsymbol{V}_r$. In Fig. 7.34, the difference between C^1 and G^1 continuity is illustrated using two composite 2nd degree Bézier splines. Also show that the sensitivities of

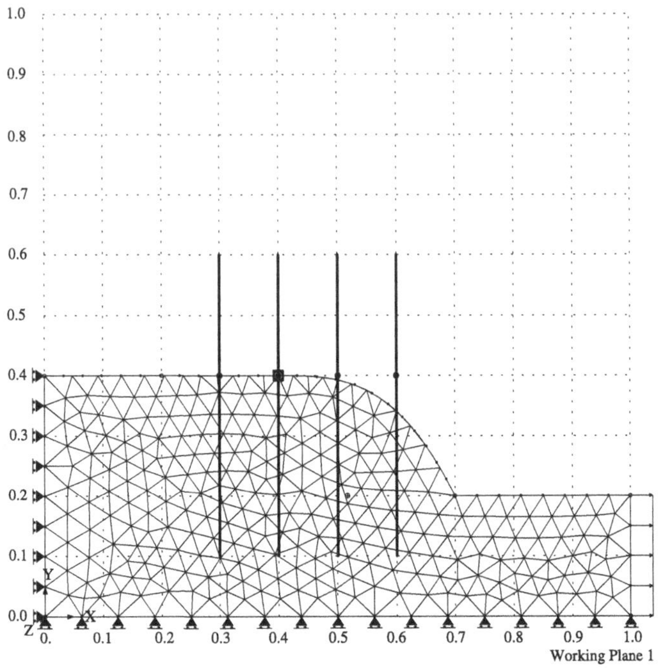

Fig. 7.30 The initial mesh together with the design spans

$\boldsymbol{V}_j$ are

$$\frac{\partial \boldsymbol{V}_j}{\partial \alpha_l} = \frac{(y_j - y_r)L_{l,x} - (x_j - x_r)L_{l,y}}{(y_l - y_r)L_{j,x} - (x_l - x_r)L_{j,y}} \boldsymbol{L}_j$$

$$\frac{\partial \boldsymbol{V}_j}{\partial \alpha_r} = \frac{(y_l - y_j)L_{r,x} - (x_l - x_j)L_{r,y}}{(y_l - y_r)L_{j,x} - (x_l - x_r)L_{j,y}} \boldsymbol{L}_j.$$

When is the denominator zero?

Exercise 7.3 A circular arc has the center point $\boldsymbol{V}_c$ and the end points $\boldsymbol{V}_s$ and $\boldsymbol{V}_e$ as shown in Fig. 7.35. A design variable α_c controls how the center point moves along the span $\boldsymbol{L}_c$. The radius of the arc should remain constant. Show that the sensitivity of $\boldsymbol{V}_s$ is given by

$$\frac{\partial \boldsymbol{V}_s}{\partial \alpha_s} = \frac{(x_c - x_s)L_{c,x} + (y_c - y_s)L_{c,y}}{(x_c - x_s)L_{s,x} + (y_c - y_s)L_{s,y}} \boldsymbol{L}_s.$$

Exercise 7.4 On the book's homepage (www.mechanics.iei.liu.se/edu_ug/strop/), you may find a computer exercise where TRINITAS should be used to optimize the shape of a bridge pillar.

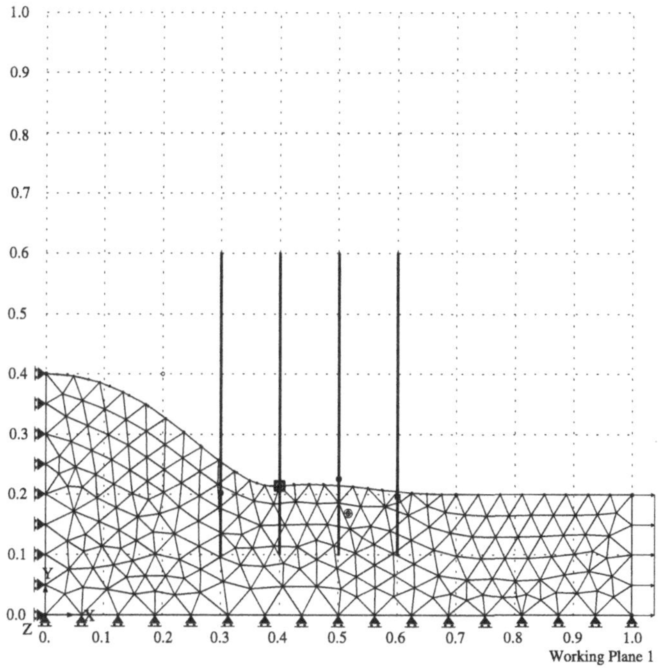

Fig. 7.31 The optimized fillet

Fig. 7.32 Two interconnected B-splines

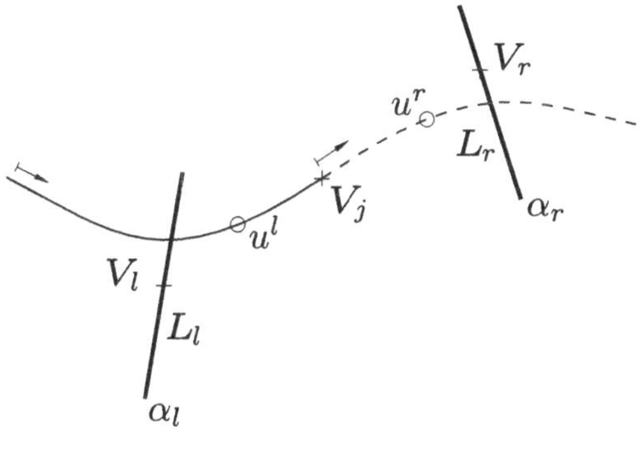

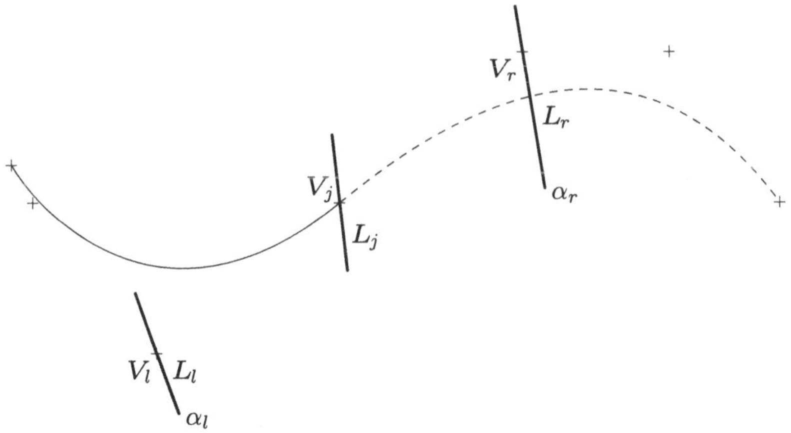

Fig. 7.33 Two interconnected Bézier splines

Fig. 7.34 Two composite curves: one C^1 continuous, the other only G^1 continuous

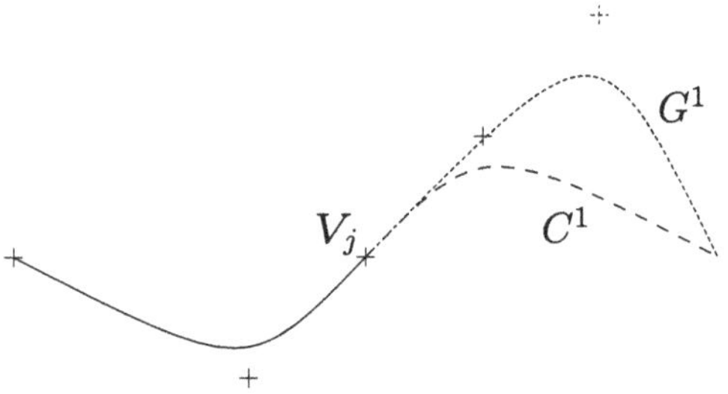

Fig. 7.35 Changing the location of the center of a circular arc

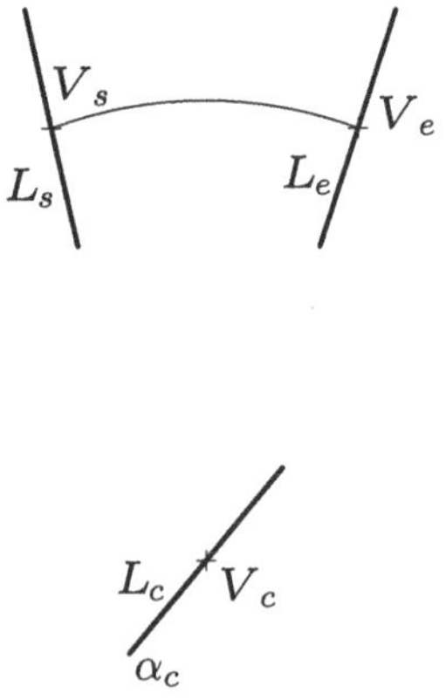

Chapter 8
Stiffness Optimization of Distributed Parameter Systems

In previous chapters we have been concerned with the solution of discrete structural optimization problems: either the structures have been naturally discrete, like trusses, or we have made them discrete by a finite element discretization. In this chapter, on the other hand, we will look at some techniques of mathematics, from an area usually referred to as calculus of variations, that can handle some continuous optimization problems such as those of distributed parameter systems, without the need for a discrete approximation. Basic facts from this area will be applied to two types of optimization problems. Firstly, we will discuss linear elastic systems without introducing any design variables. It will be shown that the state variables of such systems are minimizers of the potential energy of the systems. Next, we look at design problems of a particular type: the design variable enters linearly in the potential energy and we seek to make the structure as stiff as possible in the sense previously considered in Chap. 5. It is shown that optimal structures of this type have the property that a particular specific strain energy is constant throughout the structure, which is to be compared to the fully stressed designs of Sect. 5.2.2. We treat mainly simple problems of beams and bars, but the general structure of this stiffness optimization problem will be used in the next chapter that treats topology optimization problems.

8.1 Calculus of Variations

In elementary mathematics courses, conditions for a real valued function to take a minimum (or maximum) value are studied. It is found that at such extreme points partial derivatives are zero-valued. Calculus of variations may be seen as an extension of this elementary theory: instead of functions of several real variables, symbolized in standard notation as

$$f : \mathbb{R}^n \longrightarrow \mathbb{R},$$

calculus of variations considers *functionals*, i.e., functions of functions, for which we use the notation

$$J : D \longrightarrow \mathbb{R},$$

where D is a set of functions. We try to find the function, belonging to D, which makes J as small (or large) as possible. Thus, instead of looking for an optimal *point* in $\mathbb{R}^n$ as in elementary theory, we look for an optimal *function* in D. Usually

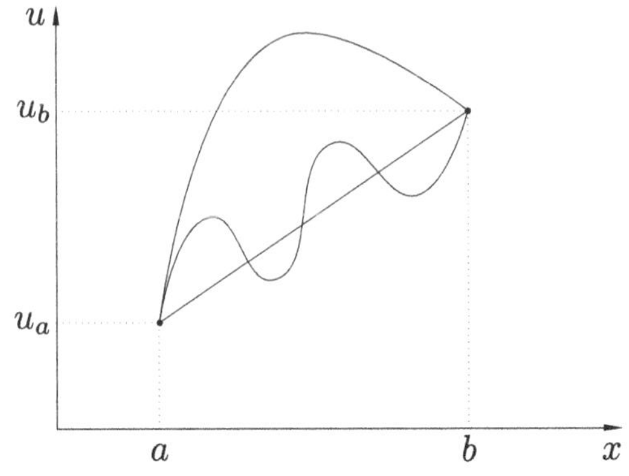

Fig. 8.1 Three different members of the set D

we need to require the functions of D to have certain properties such as a degree of smoothness. A good example of a set D, used below, is

$$D = \{u \in C^2[a,b],\ u(a) = u_a,\ u(b) = u_b\}$$

where $C^2[a,b]$ denotes all functions which are twice continuously differentiable on the closed interval $[a,b]$ of the real line, and u_a and u_b are fixed boundary values. Figure 8.1 shows how members of the set D may appear. A typical example of a functional defined on the set D is

$$J(u) = \int_a^b u\,dx.$$

In the following example we will see another such typical functional.

Finding the Shortest Path Between Points

We like to consider the problem of finding the shortest path between two points in the plane. You know that the solution of this problem is a straight line connecting the points, but do you have a proof of this? We will see that calculus of variations can provide such a proof. The problem is formulated in the frame of calculus of variations in this subsection and the actual solution is left as Exercise 8.2. Tools for doing this exercise are developed in the upcoming section.

Let the coordinates of the plane be denoted (x,u) and let the two points that should be connected be (a,u_a) and (b,u_b). Moreover, let s be the curve parameter of any curve connecting these points. We then find that

$$ds = \sqrt{dx^2 + du^2} = \sqrt{1 + (u')^2}dx,$$

where, in the second equality, we have assumed that the curve does not have any vertical parts so that u can be seen as a function of x, i.e., $u = u(x)$, and $'$ is a shorthand for the derivative of this function.[1] We have also assumed that $u(x)$ is differentiable: a sufficient condition for this is $u \in C^2[a,b]$. The length of the curve is obtained by summing (integrating) all infinitesimal parts ds, i.e.,

$$\text{Length} = \int_a^b \sqrt{1+(u')^2}dx,$$

In summary, the problem of finding the shortest path between two points is given by

$$\min_{u \in D} \int_a^b \sqrt{1+(u')^2}dx.$$

In the following we will look at techniques for solving this and similar problems.

8.1.1 Optimality Conditions and Gateaux Derivatives

We like to characterize a function $u^* \in D$ which makes a certain functional J take its minimum value, i.e., we consider the minimization problem

$$(\mathbb{P}_0) \quad \min_{u \in D} J(u)$$

and look for conditions satisfied by a solution of this problem. Suppose u^* solves $(\mathbb{P}_0)$ and let

$$u = u^* + \varepsilon\varphi,$$

be any other function in D. Here, $\varepsilon \in \mathbb{R}$ and φ is a function on the interval $[a,b]$. As can be understood from Fig. 8.2, if u should belong to D we have to require that

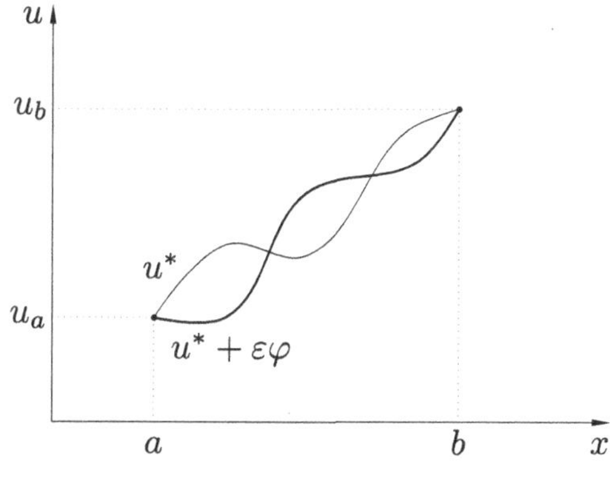

Fig. 8.2 The function u^* and the nearby function $u^* + \varepsilon\varphi$

[1] Throughout this chapter we use prime, $'$, as a shorthand for the derivative of a function of one variable.

$\varphi \in C^2[a,b]$ and $\varphi(a) = \varphi(b) = 0$. It is natural to denote $\varepsilon\varphi$ as the *variation* of u^*.

For each φ we may define the function $\phi : \mathbb{R} \longrightarrow \mathbb{R}$ as

$$\phi(\varepsilon) = J(u^* + \varepsilon\varphi).$$

By the assumption that u^* is a solution of $(\mathbb{P}_0)$, ϕ takes its minimum at $\varepsilon = 0$ and it must hold that

$$\left.\frac{d\phi(\varepsilon)}{d\varepsilon}\right|_{\varepsilon=0} = 0,$$

which, by the elementary definition of a derivative, is equivalent to

$$\lim_{\varepsilon\to 0} \frac{J(u^* + \varepsilon\varphi) - J(u^*)}{\varepsilon} = 0.$$

The quantity on the left-hand side is referred to as the *Gateaux derivative* of J at u^* in the direction φ and we use the notation

$$J'(u;\varphi) = \lim_{\varepsilon\to 0} \frac{J(u + \varepsilon\varphi) - J(u)}{\varepsilon}.$$

We have found that a *necessary* condition for u^* to solve $(\mathbb{P}_0)$ is that $J'(u^*;\varphi) = 0$ *for all* $\varphi \in C^2[a,b]$ for which $\varphi(a) = \varphi(b) = 0$. If J is *convex* it can be shown that this condition is also sufficient for u^* to solve $(\mathbb{P}_0)$. Furthermore, generalizations to slightly different situations, for instance, when $u(a) = u_a$ is excluded from the definition of the set D, will be considered in upcoming sections. The optimality condition $J'(u;\varphi) = 0$ can be compared to the condition that all partial derivatives should be zero when a function of several real variables takes an extreme value.

8.1.1.1 Examples of Gateaux Derivatives

Next, we collect a few explicit examples of Gateaux derivatives of functionals. These results will take the form of formulas to be used in subsequent sections.

1. Let $J(u) = u(x_0)$, where x_0 is any point in the interval $[a,b]$. Then one concludes that

$$J'(u;\varphi) = \lim_{\varepsilon\to 0} \frac{(u + \varepsilon\varphi)(x_0) - u(x_0)}{\varepsilon} = \varphi(x_0).$$

2. Let

$$J(u) = \int_a^b f u \, dx,$$

for some function f. One finds

$$\frac{J(u + \varepsilon\varphi) - J(u)}{\varepsilon} = \frac{\int_a^b f u\,dx + \varepsilon\int_a^b f\varphi\,dx - \int_a^b f u\,dx}{\varepsilon} = \int_a^b f\varphi\,dx$$

so it can be concluded that

$$J'(u;\varphi) = \int_a^b f\varphi\,dx.$$

3. Let

$$J(u) = \int_a^b (u')^2\,dx,$$

where $u' = du/dx$. Then

$$\begin{aligned}
&\frac{J(u+\varepsilon\varphi) - J(u)}{\varepsilon} \\
&= \frac{\int_a^b (u')^2\,dx + \varepsilon^2 \int_a^b (\varphi')^2\,dx + 2\varepsilon \int_a^b u'\varphi'\,dx - \int_a^b (u')^2\,dx}{\varepsilon} \\
&= \varepsilon \int_a^b (\varphi')^2\,dx + 2\int_a^b u'\varphi'\,dx \longrightarrow 2\int_a^b u'\varphi'\,dx \quad \text{as } \varepsilon \to 0.
\end{aligned}$$

Thus,

$$J'(u;\varphi) = 2\int_a^b u'\varphi'\,dx.$$

4. Generally it can be concluded that for

$$J(u) = \int_a^b F(u')\,dx,$$

where F is any differentiable real function, the Gateaux derivative reads

$$J'(u;\varphi) = \int_a^b F'(u')\varphi'\,dx.$$

Note that in all these four cases u and u' can be interchanged to produce analogous results. In the following we will also make use of formulas containing the second derivative u'', which follow directly by generalizations from the above. For instance, the Gateaux derivative of

$$J(u) = \int_a^b (u'')^2\,dx$$

is

$$J'(u;\varphi) = 2\int_a^b u''\varphi''\,dx.$$

8.1.1.2 Solution of a Simple Example

Consider the following problem:

$$\begin{cases} \min_{u} & \int_0^1 (u^2 + (u')^2)\,dx \\ \text{s.t.} & \begin{cases} u(0) = 0 \\ u(1) = 1 \\ u \in C^2[0, 1]. \end{cases} \end{cases}$$

According to the previous section a necessary (and, in fact, sufficient since the problem is convex) condition for u^* to be a solution of this problem can be found by studying the Gateaux derivative of

$$J(u) = \int_0^1 (u^2 + (u')^2)\,dx.$$

With straightforward generalization to u without prime, such a derivative can be calculated by using item 3 above. One finds that an optimal solution u^* satisfies

$$J'(u^*; \varphi) = \int_0^1 \left(2u^*\varphi + 2u^{*'}\varphi'\right) dx = 0, \tag{8.1}$$

for all

$$\varphi \in C^2[0, 1] \quad \text{such that } \varphi(0) = \varphi(1) = 0. \tag{8.2}$$

We integrate (8.1) by parts to obtain

$$\left[2u^{*'}\varphi\right]_0^1 + \int_0^1 \left(2u^* - 2u^{*''}\right)\varphi\,dx = 0, \tag{8.3}$$

which holds for all φ satisfying (8.2). Due to the restriction on φ at the end points of the interval, the first term of (8.3) vanishes. Once this is concluded the second term in (8.3) can be handled by means of the following lemma, known as "the fundamental lemma of the calculus of variations":

Lemma 8.1 *If f is a continuous function on $[a, b]$ and*

$$\int_a^b f\eta\,dx = 0$$

for all $\eta \in C^2[a, b]$ with $\eta(a) = \eta(b) = 0$, then $f(x) = 0$ for all $x \in [a, b]$.

The truth of this lemma should be intuitively clear and in the following sections we will allude to other similar lemmas without explicitly stating them. Applying

Lemma 8.1 to (8.3), after it has been concluded that the first term is zero, one obtains the differential equation

$$u^* - u^{*''} = 0 \quad \text{on } [0, 1],$$

which has the solution

$$u^*(x) = Ae^x + Be^{-x},$$

where A and B are constants that are determined from the boundary conditions $u^*(0) = 0$ and $u^*(1) = 1$. One then finds

$$u^*(x) = \frac{e^x - e^{-x}}{e - e^{-1}} \approx 0.42(e^x - e^{-x}) = 0.84 \sinh x.$$

The optimal value of J, i.e., $J(u^*)$, is found to be approximately 1.2796. As a control of the correctness of this result one may calculate the value of J for any other function satisfying the constraints, e.g., $u = x$. The result is

$$J(u) = \int_0^1 (x^2 + 1)\, dx = \frac{4}{3} > 1.2796.$$

8.1.2 Handling a Constraint

Consider the following problem:

$$(\mathbb{P}_1) \quad \begin{cases} \min\limits_u \ J(u) \\ \\ \text{s.t.} \quad u \in D_1 \end{cases} \iff \begin{cases} u \in C^2[a, b] \\ \\ \int_a^b f u\, dx = C, \end{cases}$$

for some given function f and constant C. The essential difference between this problem and those considered in the previous section, is a constraint in the form of an integral. A way to handle this is to form the Lagrangian function

$$\mathcal{L}(u, \lambda) = J(u) + \lambda \left(\int_a^b f u\, dx - C \right),$$

where $\lambda \in \mathbb{R}$ is a Lagrangian multiplier. The following condition is *necessary* for $u^* \in D_1$ to solve $(\mathbb{P}_1)$:

$$\mathcal{L}'(u^*, \lambda; \varphi) = J'(u; \varphi) + \lambda \int_a^b f \varphi\, dx = 0, \tag{8.4}$$

for all $\varphi \in C^2[a, b]$. If J is convex this condition is also sufficient for optimality. An essential fact that makes this conclusion possible is that the integral constraint is convex since it is linear. Equation (8.4) is a classical result of calculus of variations. Conceptually it can be seen as a KKT condition, see Sect. 3.3, for a continuous optimization problem.

8.1.2.1 Solution of an Example Including a Constraint

We will consider the problem of designing a container to include a certain volume (e.g., of water), see Fig. 8.3. The container has a height of 1 unit and also a depth of 1 unit, while the width is represented by a function $x \mapsto u(x)$; x varies from the bottom, where $x = 0$, to the top of the container, where $x = 1$. The shape is symmetric and the width for a certain height x is given by $2u(x)$.

We want to minimize the amount of material required to construct the walls of the container, not counting top and bottom. For walls of given thickness this means minimizing the area of the container walls. Letting s be a curve parameter so that

$$ds = \sqrt{dx^2 + du^2} = \sqrt{1 + (u')^2}\, dx,$$

we find that the expression to be minimized is

$$2\int 1\, ds + 2\int 2u\, dx = 2\int_0^1 \sqrt{1 + (u')^2}\, dx + 4\int_0^1 u\, dx, \tag{8.5}$$

where the number 2 in front of the integrals are included since there are four sides of the container and the 1 in the first integral represents the depth.

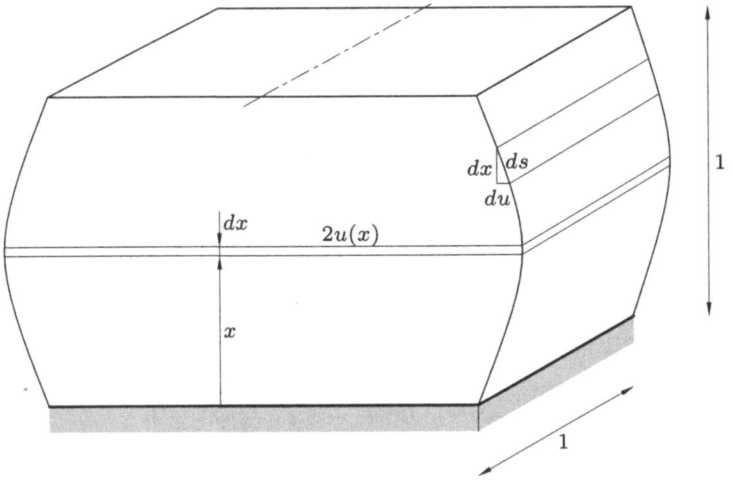

Fig. 8.3 A container of height 1, depth 1 and width $2u(x)$

The constraint saying that the container should include a certain amount of volume may be written as

$$\int_0^1 u\,dx = C,$$

for some number C. This constraint means that the last term in (8.5) is constant for all admissible designs and, thus, need not be included in the objective function. In conclusion, we are to solve the following optimization problem

$$\begin{cases} \min\limits_{u} \displaystyle\int_0^1 \sqrt{1+(u')^2}\,dx \\ \text{s.t.} \quad \begin{cases} u \in C^2[a,b] \\ \displaystyle\int_0^1 u\,dx = C. \end{cases} \end{cases}$$

We solve this problem by writing down a Lagrangian:

$$\mathcal{L}(u,\lambda) = \int_0^1 \sqrt{1+(u')^2}\,dx + \lambda\left(\int_0^1 u\,dx - C\right).$$

Using items 2 and 4 of Sect. 8.1.1.1 to calculate the Gateaux derivative, the condition for optimality, cf. (8.4), is found to be

$$\mathcal{L}'(u,\lambda;\varphi) = \int_0^1 \frac{u'\varphi'}{\sqrt{1+(u')^2}}\,dx + \lambda\int_0^1 \varphi\,dx = 0, \tag{8.6}$$

for all $\varphi \in C^2[0,1]$, where, for simplicity, we write u instead of u^*. We integrate (8.6) by parts to obtain

$$\left[\frac{u'\varphi}{\sqrt{1+(u')^2}}\right]_0^1 + \int_0^1 \left(\lambda - \left(\frac{u'}{\sqrt{1+(u')^2}}\right)'\right)\varphi\,dx = 0. \tag{8.7}$$

Equation (8.7) should be valid for all $\varphi \in C^2[0,1]$. Thus, it is also valid for all $\varphi \in C^2[0,1]$ such that $\varphi(0) = \varphi(1) = 0$. For such a restricted set of φ:s, the first term in (8.7) vanishes and Lemma 8.1 can be used to conclude that

$$\lambda = \left(\frac{u'}{\sqrt{1+(u')^2}}\right)' \quad \text{on } [0,1]. \tag{8.8}$$

Once this is concluded, (8.7) is reduced to

$$\left[\frac{u'\varphi}{\sqrt{1+(u')^2}}\right]_0^1 = 0 \quad \text{for all } \varphi(0) \text{ and } \varphi(1), \tag{8.9}$$

from which we find that

$$\frac{u'}{\sqrt{1+(u')^2}} = 0 \quad \text{at } x = 0 \text{ and } x = 1. \tag{8.10}$$

The optimal function u is now governed by (8.8) and (8.10). Integrating (8.8) we find

$$\frac{u'}{\sqrt{1+(u')^2}} = \lambda x + C_1,$$

where C_1 is an integration constant. From the boundary conditions (8.10) we conclude that $C_1 = \lambda = 0$, which means that $u' = 0$ and therefore

$$u = \text{constant} \quad \text{on } [0, 1].$$

From the integral constraint, representing the fixed volume of the container, one concludes that the constant actually has the value C.

8.2 Equilibrium Principles for Distributed Parameter Systems

This section is concerned with state problems in the form of equilibrium principles. Thus, we do not introduce any design variables that allow for modification or design of the structure. This will be the subject of the next subsection. Three different elastic distributed parameter systems will be considered. For each system we will give three, essentially equivalent, problem formulations. These formulations are

1. Potential Energy Minimization—($\mathbb{PEM}$).
2. Principle of Virtual Work—($\mathbb{PVW}$).
3. Partial Differential Equation—($\mathbb{PDE}$).

For each system it will be shown, or indicated, that

$$(\mathbb{PEM}) \quad \Longleftrightarrow \quad (\mathbb{PVW}) \quad \Longleftrightarrow \quad (\mathbb{PDE}).$$

We treat below, one-dimensional elasticity, a beam problem and two-dimensional elasticity, after which we give an abstract formulation that includes all of the previous examples, as well as many other linear elastic systems.

8.2.1 One-Dimensional Elasticity

Consider the one-dimensional rod problem of Fig. 8.4. The rod, of length L, is fixed at $x = 0$ and a force P is applied at $x = L$. The cross-sectional area A is regarded as a function of x. The total potential energy for this system may be written as

$$J(u) = \int_0^L \frac{AE}{2}(u')^2\,dx - Pu(L),$$

Fig. 8.4 A one-dimensional linear elastic system with variable area

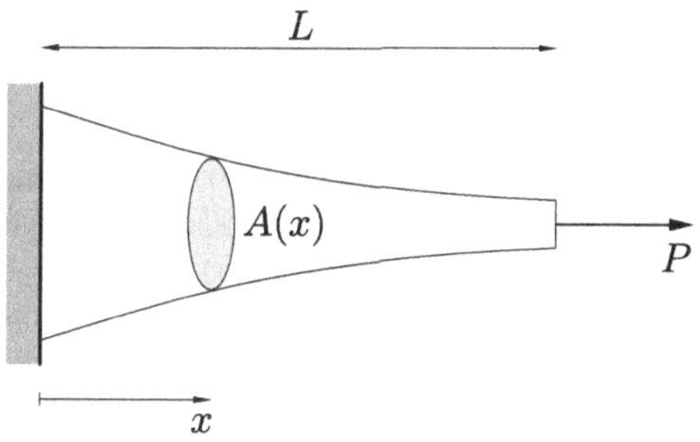

where E is the elasticity modulus and $x \mapsto u(x)$ is the displacement distribution. The first term in $J(u)$ is the strain energy and the second term is the potential of external loading.

We will consider the problem of minimizing the potential energy:

$$(\mathbb{PEM}) \quad \begin{cases} \min\limits_{u} \;\; J(u) \\ \text{s.t.} \quad \begin{cases} u \in C^2[a,b] \\ u(0) = 0. \end{cases} \end{cases}$$

The boundary condition $u(0) = 0$ comes from the rod being fixed at the left end. Since J is convex we know from the previous section that u is a solution of $(\mathbb{PEM})$ if and only if

$$J'(u;\varphi) = \int_0^L AEu'\varphi'\,dx - P\varphi(L) = 0 \tag{8.11}$$

for all

$$\varphi \in C^2[0,L] \quad \text{such that } \varphi(0) = 0. \tag{8.12}$$

The Gateaux derivative in (8.11) was calculated using items 1 and 3 of Sect. 8.1.1.1. Equation (8.11) and the admissible variations (8.12) constitute the Principle of Virtual Work $(\mathbb{PVW})$: the term $P\varphi(L)$ can be regarded as the work produced by the force P when perturbing (virtually) the displacement by the field φ; similarly, the integral term in (8.12) is the work of the stresses for such a perturbation. Next, we integrate (8.11) by parts to obtain

$$\left[AEu'\varphi\right]_0^L - \int_0^L \left(AEu'\right)'\varphi\,dx = P\varphi(L), \tag{8.13}$$

for all φ satisfying (8.12). We evaluate (8.13) by first choosing any φ which in addition to (8.12) satisfies $\varphi(L) = 0$. For such a subset of admissible φ:s, the first term of (8.13) vanishes and Lemma 8.1 gives

$$\left(AEu'\right)' = 0 \quad \text{on } [0,L]. \tag{8.14}$$

This result is substituted into (8.13) which makes the second term disappear. By returning to the admissible φ:s of (8.12) we conclude from the remaining term of

(8.13) that

$$AEu' = P \quad \text{at } x = L. \tag{8.15}$$

In conclusion, a displacement distribution u that makes the potential energy take a minimum value, i.e., that solves ($\mathbb{PEM}$), is governed by the differential equation (8.14), the boundary conditions (8.15) and $u(0) = 0$. The formulation given by these equations is called the Partial Differential Equation, ($\mathbb{PDE}$), formulation. We recognize this formulation as one that governs *equilibrium* of the rod and we may say that minimizing the potential energy means finding an equilibrium displacement.

Note how the two boundary conditions are introduced in different ways when the problem is seen as a minimization problem: the condition $u(0) = 0$ is a part of the definition of the competing, or admissible, displacements, while (8.15) is an optimality condition resulting from making the Gateaux derivative zero. There is a terminology for these two types of boundary conditions: $u(0) = 0$ is called an *essential* boundary condition, while (8.15) is a *natural* boundary condition.

8.2.2 Beam Problem

Consider the Euler-Bernoulli beam of Fig. 8.5. The beam is of length L and is built-in at both ends so the boundary conditions on transverse displacement $x \mapsto u(x)$ and its derivative are $u(0) = u(L) = u'(0) = u'(L) = 0$. The area moment of inertia I may be a function of x. The beam is acted on by a possibly nonconstant transverse force per unit length q. The total potential energy for this system may be written as

$$J(u) = \int_0^L \frac{EI}{2}(u'')^2\,dx - \int_0^L qu\,dx,$$

where E is the elasticity modulus. As usual, the first term in $J(u)$ is the strain energy and the second term is the potential of external loading.

We will consider the problem of minimizing the potential energy:

$$(\mathbb{PEM}) \quad \begin{cases} \min\limits_{u} \; J(u) \\ \text{s.t.} \;\; u(0) = u(L) = u'(0) = u'(L) = 0, \end{cases}$$

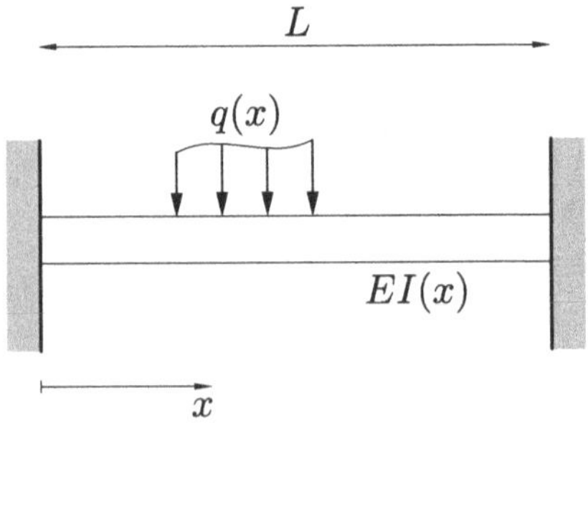

Fig. 8.5 A built-in beam

where, for simplicity, we have not explicitly expressed the degree of smoothness required for u.

We know that u is a solution of ($\mathbb{PEM}$) if and only if

$$J'(u;\varphi) = \int_0^L EIu''\varphi''\,dx - \int_0^L q\varphi\,dx = 0 \tag{8.16}$$

for all

$$\varphi \text{ such that } \quad \varphi(0) = \varphi(L) = \varphi'(0) = \varphi'(L) = 0. \tag{8.17}$$

The Gateaux derivative in (8.16) was calculated using items 2 and 3 of Sect. 8.1.1.1. Equation (8.16) and the admissible variations (8.17) constitute the Principle of Virtual Work ($\mathbb{PVW}$). We integrate (8.16) twice by parts: the first integration gives

$$\left[EIu''\varphi'\right]_0^L - \int_0^L \left(EIu''\right)'\varphi'\,dx = \int_0^L q\varphi\,dx,$$

where the first term vanishes due to (8.17), and the second integration gives

$$\left[-(EIu'')\varphi\right]_0^L + \int_0^L \left[\left(EIu''\right)'' - q\right]\varphi\,dx = 0, \tag{8.18}$$

where, again, the first term vanishes due to (8.17). Now, Lemma 8.1 gives

$$\left(EIu''\right)'' = q \quad \text{on } [0, L], \tag{8.19}$$

which is the Euler-Bernoulli beam deflection equation. Equation (8.19), together with boundary conditions, represent the ($\mathbb{PDE}$) formulation for this system. Note that all boundary conditions are essential such conditions in this example.

8.2.3 Two-Dimensional Elasticity

Consider a two-dimensional domain $\Omega \in \mathbb{R}^2$ and let h be a thickness. The three-dimensional domain $\Omega \times [0, h]$ is occupied by a linear elastic body, a sheet, which is loaded, and also deforms, in the plane containing Ω. The boundary of Ω is decomposed into the two parts, Γ_t and Γ_u, and a point in Ω is denoted $\boldsymbol{x} = (x, y)$. External forces acting on the body are force per unit area $\boldsymbol{b}(\boldsymbol{x}) \in \mathbb{R}^2$, defined for $\boldsymbol{x} \in \Omega$, and force per unit length $\boldsymbol{t}(\boldsymbol{x}) \in \mathbb{R}^2$, defined for $\boldsymbol{x} \in \Gamma_t$. The domain Ω, its boundaries, and the loadings are shown in Fig. 8.6. Internal forces of the sheet can be represented by the normal stresses σ_x and σ_y and the shear stress σ_{xy}. A column vector is formed from these stresses:

$$\boldsymbol{\sigma} = \left[\sigma_x, \sigma_y, \sigma_{xy}\right]^T.$$

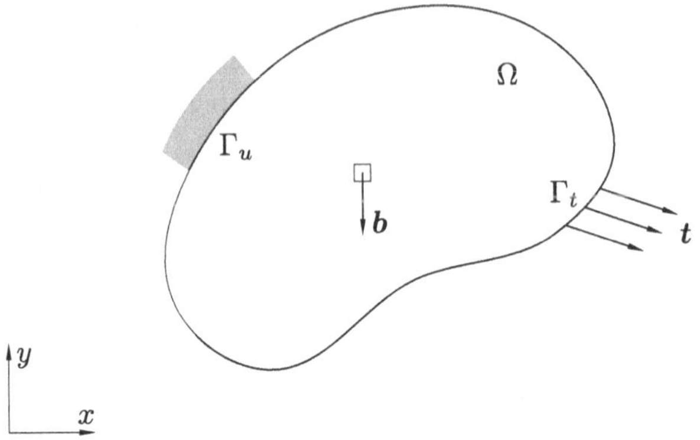

Fig. 8.6 The two-dimensional elastic domain Ω

We want to write down the equations governing equilibrium for the linear elastic body. To that end, we use a notational scheme that can be found in, e.g., Ottosen and Petersson [24]. We define an operator matrix as

$$\boldsymbol{\nabla}^T = \begin{bmatrix} \dfrac{\partial}{\partial x} & 0 & \dfrac{\partial}{\partial y} \\ 0 & \dfrac{\partial}{\partial y} & \dfrac{\partial}{\partial x} \end{bmatrix}$$

and equilibrium then reads

$$\boldsymbol{\nabla}^T(h\boldsymbol{\sigma}) + \boldsymbol{b} = \boldsymbol{0} \quad \text{on } \Omega. \tag{8.20}$$

On the boundary part Γ_t, there must be equilibrium between the internal forces represented by $\boldsymbol{\sigma}$ and the external boundary force $\boldsymbol{t}$. This may be formulated by introducing the matrix

$$\boldsymbol{N} = \begin{bmatrix} n_x & 0 & n_y \\ 0 & n_y & n_x \end{bmatrix}$$

where $\boldsymbol{n} = (n_x, n_y)$ is the unit outward normal vector of Γ_t. Boundary equilibrium then reads

$$\boldsymbol{N}(h\boldsymbol{\sigma}) = \boldsymbol{t} \quad \text{on } \Gamma_t. \tag{8.21}$$

This equation reads in components

$$t_x = (h\sigma_x)n_x + (h\sigma_{xy})n_y$$
$$t_y = (h\sigma_{xy})n_x + (h\sigma_y)n_y.$$

The displacement is a vector $\boldsymbol{u} = [u_x, u_y]^T$. On Γ_u the body is fixed so this displacement vector is zero, i.e.,

$$\boldsymbol{u} = \boldsymbol{0} \quad \text{on } \Gamma_u. \tag{8.22}$$

What remains to be specified is the constitutive equation for the linear elastic body. Such an equation is written in terms of strains, which can be collected in a column vector $\boldsymbol{\varepsilon}$ and are defined by

$$\boldsymbol{\varepsilon} = \nabla \boldsymbol{u}. \tag{8.23}$$

The constitutive law of linear elasticity, usually known as Hooke's law, now reads

$$\boldsymbol{\sigma} = \boldsymbol{D}\boldsymbol{\varepsilon}, \tag{8.24}$$

where $\boldsymbol{D}$ is the constitutive matrix. For an isotropic material, assuming plane stress conditions, this matrix was introduced already on page 106 and reads

$$\boldsymbol{D} = \frac{E}{1-\nu^2}\begin{bmatrix} 1 & \nu & 0 \\ \nu & 1 & 0 \\ 0 & 0 & \dfrac{1-\nu}{2} \end{bmatrix}, \tag{8.25}$$

where E is Young's modulus and ν is Poisson's ratio.

In summary, the problem of two-dimensional linear elasticity is governed by Eqs. (8.20), (8.21), (8.22), (8.23) and (8.24). We eliminate $\boldsymbol{\sigma}$ and $\boldsymbol{\varepsilon}$ from these equations to obtain the Partial Differential Equation formulation, which is to find $\boldsymbol{u} : \Omega \to \mathbb{R}^2$ such that

$$(\mathbb{PDE}) \quad \begin{cases} \nabla^T(h\boldsymbol{D}\nabla\boldsymbol{u}) + \boldsymbol{b} = \boldsymbol{0} & \text{on } \Omega \\ \boldsymbol{N}(h\boldsymbol{D}\nabla\boldsymbol{u}) = \boldsymbol{t} & \text{on } \Gamma_t \\ \boldsymbol{u} = \boldsymbol{0} & \text{on } \Gamma_u. \end{cases}$$

Following a standard procedure, involving a two-dimensional version of Lemma 8.1, it can be shown that for sufficiently smooth displacement fields, the $(\mathbb{PDE})$ formulation is equivalent to the Principle of Virtual Work:

$$(\mathbb{PVW}) \quad \begin{cases} \text{Find } \boldsymbol{u} : \Omega \to \mathbb{R}^2 \text{ such that } \boldsymbol{u} = \boldsymbol{0} \text{ on } \Gamma_u \text{ and} \\ \displaystyle\int_\Omega (\nabla\boldsymbol{v})^T h\boldsymbol{D}\nabla\boldsymbol{u}\, dA = \int_{\Gamma_t} \boldsymbol{v}^T\boldsymbol{t}\, ds + \int_\Omega \boldsymbol{v}^T\boldsymbol{b}\, dA \\ \text{for all } \boldsymbol{v} : \Omega \to \mathbb{R}^2 \text{ that equals zero on } \Gamma_u. \end{cases}$$

Finally, we consider the total potential energy

$$J(\boldsymbol{u}) = \frac{1}{2}\int_\Omega (\nabla\boldsymbol{u})^T h\boldsymbol{D}\nabla\boldsymbol{u}\, dA - \int_{\Gamma_t} \boldsymbol{u}^T\boldsymbol{t}\, ds - \int_\Omega \boldsymbol{u}^T\boldsymbol{b}\, dA,$$

where, as usual, the first term is the strain energy and the second two terms are the potential of external loading. We consider the minimization of potential energy over admissible displacement fields:

$$(\mathbb{PEM})\quad \begin{cases} \min\limits_{u} \; J(\boldsymbol{u}) \\ \text{s.t.} \quad \boldsymbol{u} = \boldsymbol{0} \quad \text{on } \Gamma_u. \end{cases}$$

This is a convex problem and the optimality condition becomes

$$J'(\boldsymbol{u}; \boldsymbol{v}) = 0 \quad \text{for all } \boldsymbol{v} : \Omega \to \mathbb{R}^2 \text{ that equals zero on } \Gamma_u \quad \Longleftrightarrow \quad (\mathbb{PVW}),$$

and so we have equivalence between our three problems.

Note that ($\mathbb{PDE}$) contains second order derivatives with respect to the displacement, while the other two formulations contain only first order derivatives. Thus, ($\mathbb{PEM}$) and ($\mathbb{PVW}$) make sense for a larger class of displacement fields than ($\mathbb{PDE}$) and, therefore, the latter formulation is called a *strong* formulation, while the other two are *weak* formulations.

8.2.4 Abstract Equilibrium Principles

The three distributed parameter systems of previous subsections can be covered by one abstract formulation. To that end, the total potential energy is written as

$$J(\mathcal{V}) = \frac{1}{2} a(\mathcal{V}, \mathcal{V}) - \ell(\mathcal{V}),$$

where

- $\mathcal{V}$ belongs to a set K of admissible displacement fields, i.e., fields that are sufficiently regular and satisfy necessary, or kinematic, boundary conditions,
- a is a symmetric bilinear functional, i.e., it satisfies $a(\mathcal{U}, \mathcal{V}) = a(\mathcal{V}, \mathcal{U})$ for all $\mathcal{V}, \mathcal{U} \in K$ and it is linear in both of its arguments, and
- ℓ is a linear functional.

Comparing with previous subsections, $\mathcal{V} \in K$ should be identified with the longitudinal displacement of a rod, denoted u, transverse displacement for a beam, also denoted u, or the displacement vector of a sheet, denoted $\boldsymbol{u}$.

The problem of Potential Energy Minimization can abstractly be written

$$(\mathbb{PEM}) \quad \text{Find } \mathcal{U} \in K \text{ such that} \quad J(\mathcal{U}) \leq J(\mathcal{V}) \quad \text{for all } \mathcal{V} \in K.$$

An alternative statement is

$$(\mathbb{PEM}) \quad \text{Find } \mathcal{U} \in K \text{ such that} \quad J(\mathcal{U}) = \min_{\mathcal{V} \in K} J(\mathcal{V}).$$

The properties of a and ℓ give that the Gateaux derivative of $J(\mathcal{U})$ becomes

$$J'(\mathcal{U}; \mathcal{V}) = \lim_{\varepsilon \to 0} \frac{J(\mathcal{U} + \varepsilon \mathcal{V}) - J(\mathcal{U})}{\varepsilon} = a(\mathcal{U}, \mathcal{V}) - \ell(\mathcal{V}).$$

The Principle of Virtual Work, which is equivalent to ($\mathbb{PEM}$), then follows by requiring that the Gateaux derivative is zero for all admissible variations. We obtain

$$(\mathbb{PVW}) \quad \text{Find } \mathcal{U} \in K \text{ such that } \quad a(\mathcal{U}, \mathcal{V}) = \ell(\mathcal{V}) \quad \text{for all } \mathcal{V} \in K,$$

where it has been assumed, for simplicity, that both the solution function $\mathcal{U}$ and the variation $\mathcal{V}$ belong to the same set K.

An identity that will be used in the next section follows by taking the arbitrary function $\mathcal{V} \in K$ to be the solution $\mathcal{U}$ in ($\mathbb{PVW}$):

$$a(\mathcal{U}, \mathcal{U}) = \ell(\mathcal{U}). \tag{8.26}$$

This equality is known as Clapeyron's theorem in applied elasticity texts.

8.3 The Design Problem

With reference to the abstract equilibrium principle introduced in the last paragraph of the last section, we will consider design problems where a design function, denoted by ρ, enters linearly in the bilinear form a. It will be assumed that the bilinear form, now considered as a function of ρ as well as of the displacement field $\mathcal{V}$, can be written

$$a(\rho, \mathcal{V}, \mathcal{V}) = 2 \int_{\Omega} \rho e(\mathcal{V})\, d\Omega. \tag{8.27}$$

Here, e is the specific strain energy, and the integration domain Ω and the integration element $d\Omega$ are made concrete in each special application of the abstract theory. The specific strain energy is a quadratic function of the displacement. If $\rho\, d\Omega$ has the physical dimension of volume, which is the case in most applications, then the term *specific* means *per volume.*

The fact that ρ enters linearly in a is assumed since, as will be indicated in Chap. 9, if a nonlinear dependence is present there will be difficulties related to nonexistence of solutions of the corresponding design problem. These difficulties can be fixed by introducing restrictions, but involve technicalities that we refrain from discussing until later. Furthermore, as will also be understood later, a linear dependence gives a convex problem, but a nonlinear dependence usually leads to a nonconvex problem.

Our design goal is to make the structure as stiff as possible. To that end we use as objective function the value of the linear form ℓ for the equilibrium displacement field $\mathcal{U}$, i.e., $\ell(\mathcal{U})$. This measure is called *compliance* and is the inverse of a global stiffness, so when we minimize compliance the structure is made as stiff as possible. From (8.26) we conclude that compliance is also equal to the bilinear form a when evaluated for the equilibrium displacement. In this context, see also Chap. 5, where compliance was used as objective when optimizing truss structures.

The designs that are allowed to compete during the design optimization are those that satisfy

$$\int_{\Omega} \rho \, d\Omega = V, \tag{8.28}$$

for a given constant V. In all applications of the general theory that we will see in this text, this constraint can be regarded as a constraint on available volume. The constant V will mostly, but not always, have the physical dimension of volume. Furthermore, the particular examples presented below show that, for physical reasons, the following additional constraint can always be enforced:

$$\rho(x) \geq 0 \quad \text{for all } x \in \Omega. \tag{8.29}$$

Summarizing, we have the following optimal design problem:

$$(\mathbb{P}_s) \quad \begin{cases} \min\limits_{\mathcal{U},\rho} \ \ell(\mathcal{U}) \\ \text{s.t.} \quad \begin{cases} \mathcal{U} \in K \text{ such that} \quad a(\rho,\mathcal{U},\mathcal{V}) = \ell(\mathcal{V}) \quad \text{for all } \mathcal{V} \in K \\ \displaystyle\int_{\Omega} \rho \, d\Omega = V \\ \rho(x) \geq 0 \quad \text{for all } x \in \Omega. \end{cases} \end{cases}$$

The first constraint is the equilibrium constraint which we can also write as the minimization problem

$$\text{Find } \mathcal{U} \in K \text{ such that} \quad J(\rho,\mathcal{U}) \leq J(\rho,\mathcal{V}) \quad \text{for all } \mathcal{V} \in K,$$

where

$$J(\rho,\mathcal{V}) = \frac{1}{2} a(\rho,\mathcal{V},\mathcal{V}) - \ell(\mathcal{V}).$$

The second two constraints of $(\mathbb{P}_s)$ are design constraints. The set of functions ρ satisfying these constraints is denoted $\mathcal{H}$, i.e.,

$$\begin{cases} \displaystyle\int_{\Omega} \rho \, d\Omega = V \\ \rho(x) \geq 0 \quad \text{for all } x \in \Omega. \end{cases} \iff \quad \rho \in \mathcal{H}.$$

We will now give a few explicit examples, related to the Sects. 8.2.1 through 8.2.3, where this general theory fits:

Example: One-dimensional elasticity

This example relates to the equilibrium problem in Sect. 8.2.1. For a rod or bar we may identify $\mathcal{V}$ with the longitudinal displacement v and the design function ρ with the cross-sectional area A to obtain

$$a(\rho, \mathcal{V}, \mathcal{V}) = \int_0^L E(v')^2 \rho \, dx.$$

The specific strain energy becomes

$$e(\mathcal{V}) = \frac{1}{2} E(v')^2,$$

and the volume constraint can be written

$$\int_0^L \rho \, dx = V.$$

Example: Beam with variable width

This example relates to Sect. 8.2.2. Consider a beam with rectangular cross section. Such a beam has an area moment of inertia

$$I = \frac{bh^3}{12},$$

where b is the width and h is the height of the cross section. If we want the design to enter linearly in the bilinear form, ρ may be identified with b. Identifying $\mathcal{V}$ with the vertical displacement v we may write

$$a(\rho, \mathcal{V}, \mathcal{V}) = \int_0^L \frac{Eh^3}{12} (v'')^2 \rho \, dx.$$

The specific strain energy becomes

$$e(\mathcal{V}) = \frac{Eh^3}{24} (v'')^2.$$

The constraint that the total volume of the beam should have a value C can be reformulated to match the general form (8.28) as follows:

$$\int_0^L h\rho \, dx = C \quad \Longleftrightarrow \quad \int_0^L \rho \, dx = \frac{C}{h} = V$$

so V has the physical dimension of area in this case.

Example: Two-dimensional elasticity

This example relates to Sect. 8.2.3 and will be central in the next chapter when we introduce topology optimization methods. The displacement field $\mathcal{V}$ is identified with the displacement vector $\boldsymbol{v}$ and ρ is the thickness h. We may write

$$a(\rho,\mathcal{V},\mathcal{V}) = \int_{\Omega} (\nabla \boldsymbol{v})^T \rho \boldsymbol{D} \nabla \boldsymbol{v} \, dA.$$

The specific strain energy becomes

$$e(\mathcal{V}) = \frac{1}{2}(\nabla \boldsymbol{v})^T \boldsymbol{D} \nabla \boldsymbol{v},$$

and the volume constraint is

$$\int_0^L \rho \, dA = V.$$

8.3.1 Optimality Conditions

In this section, we derive optimality conditions for problem $(\mathbb{P}_s)$, i.e., a system of equations that characterize a solution of this optimization problem. To that end we can use tools from calculus of variations such as Gateaux derivatives. However, we first write $(\mathbb{P}_s)$ in a nested form, i.e., we eliminate the displacement variable $\mathcal{U}$. Due to the particular form of $(\mathbb{P}_s)$ we can do this without really solving the equilibrium problem and it turns out that the nested formulation is related to a max-min saddle point problem.

For each $\rho \in \mathcal{H}$ we consider those $\mathcal{U}$ that are corresponding equilibrium solutions, here denoted $\mathcal{U}_\rho$. We may use (8.26) to reformulate the objective function as follows:

$$\begin{aligned} \ell(\mathcal{U}_\rho) &= 2\ell(\mathcal{U}_\rho) - \ell(\mathcal{U}_\rho) = 2\ell(\mathcal{U}_\rho) - a(\rho,\mathcal{U}_\rho,\mathcal{U}_\rho) \\ &= -2J(\rho,\mathcal{U}_\rho) = -2\min_{\mathcal{V}\in K} J(\rho,\mathcal{V}), \end{aligned}$$

where the last two steps used the definition of the total potential energy and the fact that $(\mathbb{PEM})$ implies that

$$J(\rho,\mathcal{U}_\rho) = \min_{\mathcal{V}\in K} J(\rho,\mathcal{V}).$$

Inserting this rewritten objective function into $(\mathbb{P}_s)$ and removing the, now always satisfied, equilibrium condition we get the following problem:

$$(\mathbb{P}_s)^{\text{alt}} \quad \max_{\rho\in\mathcal{H}} \phi(\rho),$$

where

$$\phi(\rho) = \min_{\mathcal{V}\in K} J(\rho,\mathcal{V}).$$

Since $\mathcal{U}_\rho$ is not present in this rewriting, we have a nested formulation of $(\mathbb{P}_s)$.

The max-min character of $(\mathbb{P}_s)^{\text{alt}}$ hints at a saddle point formulation. In fact, a solution ρ^* of $(\mathbb{P}_s)^{\text{alt}}$ is also the ρ-part of the solution of the following saddle point problem:

$$(\mathbb{SP}_s) \quad \begin{cases} \text{Find } \mathcal{U} \in K \text{ and } \rho^* \in \mathcal{H} \text{ such that} \\ J(\rho,\mathcal{U}) \le J(\rho^*,\mathcal{U}) \le J(\rho^*,\mathcal{V}) \\ \text{for all } \mathcal{V} \in K \text{ and } \rho \in \mathcal{H}. \end{cases}$$

A solution of $(\mathbb{SP}_s)$ is such that $\mathcal{U}$ is an equilibrium solution for ρ^*, i.e., the right-hand inequality holds, and, simultaneously, ρ^* maximizes J for $\mathcal{U}$, i.e., the left-hand inequality holds.

The connection between $(\mathbb{P}_s)^{\text{alt}}$ and $(\mathbb{SP}_s)$ follows since $J(\rho,\mathcal{U})$ is a convex-concave function: it is quadratic and convex in the second argument and linear and, thus, concave in the first argument. In the next chapter we will look at similar problems where the functional is nonconcave in the first argument.

Optimality conditions for the saddle point problem, and thus for $(\mathbb{P}_s)$, can be found by considering stationary points for the Lagrangian function

$$\mathcal{L}(\rho,\mathcal{U},\lambda) = J(\rho,\mathcal{U}) + \lambda\left(\int_\Omega \rho \, d\Omega - V\right),$$

where, considering (8.27), we have

$$J(\rho,\mathcal{U}) = \frac{1}{2}\int_\Omega \rho e(\mathcal{U}) \, d\Omega - \ell(\mathcal{U}).$$

A condition for optimality becomes

$$\mathcal{L}'(\rho,\mathcal{U},\lambda;\varphi) = \int_\Omega \varphi e(\mathcal{U}) \, d\Omega + \lambda \int_\Omega \varphi \, d\Omega = 0,$$

for all functions φ defined on Ω. This is rewritten as

$$\int_\Omega \varphi(\lambda + e(\mathcal{U})) \, d\Omega = 0,$$

for all φ. By applying Lemma 8.1, or rather a generalization of this lemma since Ω may be more general than an interval of the real line, we find

$$e(\mathcal{U}) = -\lambda = \text{constant} \quad \text{in } \Omega. \tag{8.30}$$

We may summarize the result as a theorem:

Theorem 8.1 *Assuming that the constraint* (8.29) *is not active, the functions* $\mathcal{U} \in K$ *and* $\rho^* \in \mathcal{H}$ *solve the maximum stiffness problem* $(\mathbb{P}_s)$ *if the specific strain energy is constant in* Ω, *i.e, equation* (8.30) *holds, and* $\mathcal{U}$ *is an equilibrium displacement for the design* ρ^*.

If ρ is allowed to take the value zero somewhere in Ω, the contents of Theorem 8.1 still holds but the optimality condition (8.30) needs to be replaced by

$$e(\mathcal{U}) = \begin{cases} \text{constant} & \text{on the part of } \Omega \text{ where } \rho > 0 \\ \text{any function} & \text{on the part of } \Omega \text{ where } \rho = 0. \end{cases} \tag{8.31}$$

8.3.2 The Stiffest Rod

Consider the one-dimensional rod problem of Fig. 8.7. The rod, of length 1, is fixed at $x = 0$ and a force per unit length, $b > 0$, is applied. The cross-sectional area ρ is regarded as a function of x. We want to find the distribution of this area that gives the stiffest structure. The available material volume is V and therefore

$$\int_0^1 \rho(x)\, dx = V. \tag{8.32}$$

The total potential energy for this system may be written as

$$J(\rho, v) = \frac{1}{2}\int_0^1 E(v')^2 \rho\, dx - \int_0^1 bv\, dx,$$

where E is the elasticity modulus and the specific strain energy is

$$e(\mathcal{V}) = \frac{1}{2}E(v')^2.$$

The general theory has shown that the rod is maximally stiff when the specific strain energy is constant, i.e., when

$$E(u')^2 = \lambda \quad \Leftrightarrow \quad |u'| = \sqrt{\frac{\lambda}{E}} \tag{8.33}$$

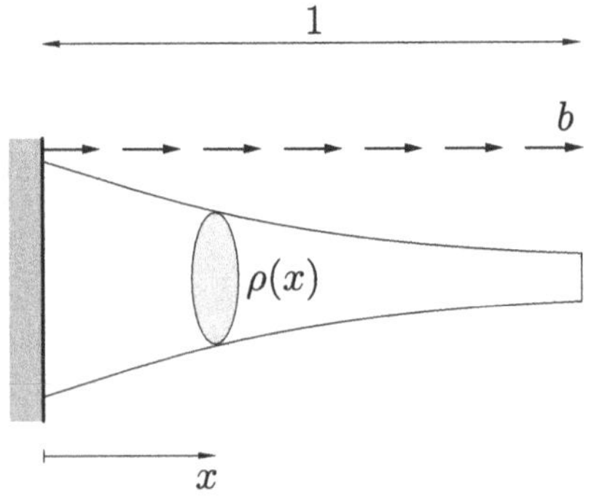

Fig. 8.7 A rod for which we like to find the optimal cross section distribution $\rho(x)$

for a constant λ which must be positive, and for the equilibrium displacement u.

Solutions to the optimal stiffness problem satisfy (8.32) and (8.33) together with equilibrium constraints on the displacement. These are found by minimizing the potential energy, i.e., by investigating the condition that the Gateaux derivative of the potential energy is zero:

$$J'(\rho,u;\varphi)=\int_0^1 \rho E u'\varphi'\,dx-\int_0^1 b\varphi'\,dx=0$$

for all φ such that $\varphi(0)=0$. Integrating this expression by parts gives

$$\left[\rho E u'\varphi\right]_0^1-\int_0^1\left[(\rho E u')'+b\right]dx=0.$$

By using Lemma 8.1 and particular choices for the test function φ we find that

$$(\rho E u')'+b=0 \quad \text{on } (0,1) \tag{8.34}$$

and

$$\rho E u'=0 \quad \text{at } x=1. \tag{8.35}$$

An optimal solution is now fully characterized by (8.32), (8.33), (8.34) and (8.35) together with the boundary condition $u(0)=0$. These equations are solved as follows: From (8.34) one concludes that

$$\rho u'=-\frac{b}{E}x+C_1,$$

for some constant C_1. The boundary condition (8.35) implies that this constant is equal to $-b/E$ so we know that

$$\rho u'=\frac{b}{E}(1-x) \quad \text{on } (0,1).$$

Since we can conclude that the right-hand side of this expression is nonnegative and so is also ρ, it is found that

$$\rho|u'|=\frac{b}{E}(1-x)$$

and by (8.33) one finds

$$\rho=C_2(1-x),$$

where C_2 is a constant. The value of this constant is found by substituting into (8.32), which gives $C_2=2V$, so our optimal cross section distribution is the wedge shape

$$\rho=2V(1-x).$$

8.3.3 Beam Stiffness Optimization

Before optimizing the distribution of material along the length of a beam we will consider the shape of an isolated cross section. In any beam problem the magnitude of the displacement is proportional to the inverse of the area moment of inertia I. With reference to Fig. 8.8, the moment of inertia for bending in the z-direction is

$$I = \int_A z^2 \, dA.$$

Given a fixed area measure we want to maximize I by changing the shape of the cross section. To that end we introduce two design variables, x_1 and x_2, according to Fig. 8.9. The following problem is now considered, based on the idea that we

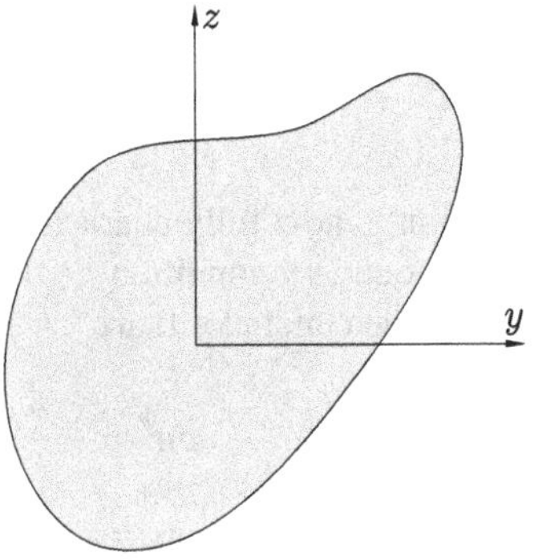

Fig. 8.8 Cross-sectional area

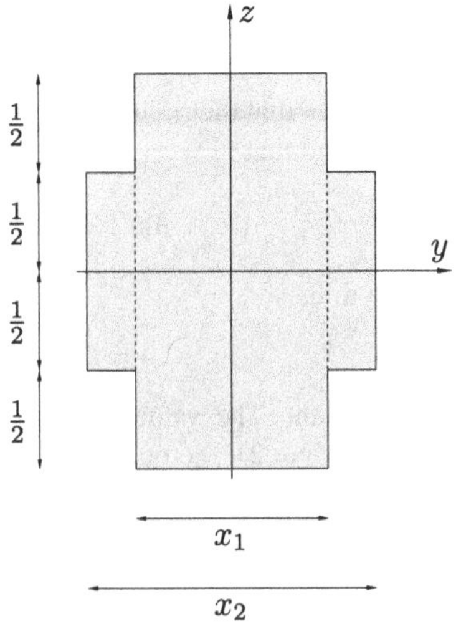

Fig. 8.9 Definition of design variables

want to minimize the beam displacement:

$$\begin{cases} \max\limits_{x_1,x_2} \; I(x_1,x_2) \\ \text{s.t.} \quad \begin{cases} \varepsilon \le x_i \le 1 & (i=1,2) \\ x_1 + x_2 = 1 & \text{(given area)} \end{cases} \end{cases}$$

where ε is a small given number. One finds

$$I(x_1,x_2) = 2\int_0^{1/2} x_2 z^2 \, dz + 2\int_{1/2}^{1} x_1 z^2 \, dz = \frac{1}{12}x_2 + \frac{7}{12}x_1.$$

The constraint of given area, $x_2 = 1 - x_1$, implies that

$$I(x_1,x_2) = \frac{1}{12} + \frac{x_1}{2},$$

so our problem now becomes

$$\begin{cases} \max\limits_{x_1} \; x_1 \\ \text{s.t.} \quad \begin{cases} \varepsilon \le x_1 \le 1 \\ \varepsilon \le 1 - x_1 \le 1 \quad \Leftrightarrow \quad [x_1 \le 1-\varepsilon, x_1 \ge 0]. \end{cases} \end{cases}$$

Clearly the solution is $x_1^* = 1 - \varepsilon$ which implies $x_2^* = \varepsilon$. Geometrically this is an I-beam as shown in Fig. 8.10. The optimum value of the bending moment of inertia is

$$I(x_1^*, x_2^*) = \frac{\varepsilon}{24} + \frac{7}{24}(1-\varepsilon) \to \frac{7}{24} \quad \text{as } \varepsilon \to 0.$$

Clearly, the optimum cross section is achieved when as much material as possible is placed in the flanges: in practice, the waist keeps the flanges in place and carries some shear stress, but almost all bending stiffness comes from the flanges.

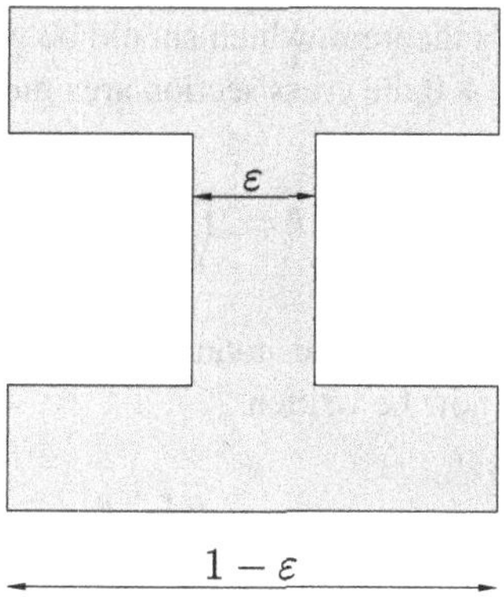

Fig. 8.10 Optimal beam cross section in the form of an I-beam

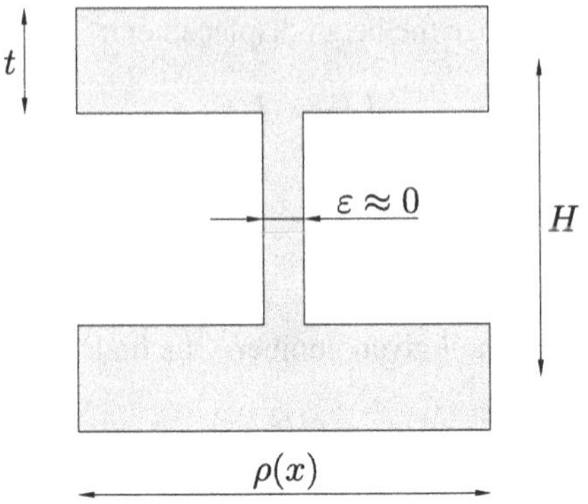

Fig. 8.11 Beam cross section shape used in the stiffness optimization problem represented in Fig. 8.12

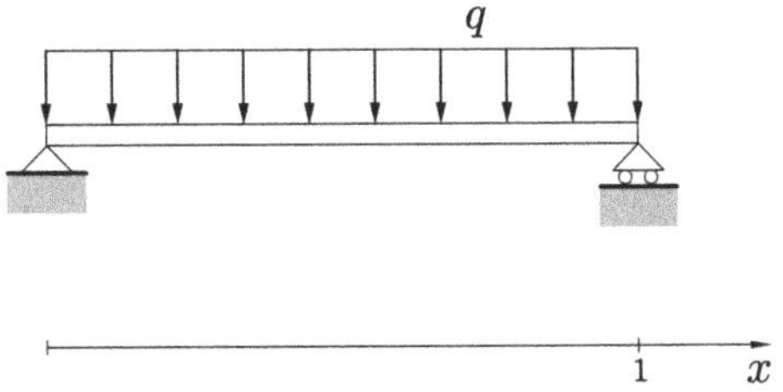

Fig. 8.12 Beam to be optimized

We now return to the stiffness optimization problem of choosing an optimal function $\rho(x)$. Since the I-beam seems to be a good cross section shape we choose this and let $\rho(x)$ represent its width, as shown in Fig. 8.11. In this figure, ε is a small number, t and H are fixed and $\rho(x)$ varies along the length of the beam. Fig. 8.12 shows the beam to be optimized: it is hinged at the ends, of length 1 and is acted on by a constant load per unit length q. The constraint that a fixed volume of material V is available is written

$$\int_0^1 2\rho t\, dx = V. \tag{8.36}$$

From Steiner's theorem, which should be well known from elementary courses, we find that as $\varepsilon \to 0$ the cross section area moment of inertia becomes

$$I = 2\left(\frac{t^3\rho}{12} + \left(\frac{H}{2}\right)^2 \rho t\right) = I_0\rho,$$

where we introduced the notation I_0. The total potential energy of the beam in Fig. 8.12 can now be written

$$J(\rho, v) = \frac{EI_0}{2}\int_0^1 \rho(v'')^2\, dx - \int_0^1 qv\, dx,$$

where, as usual, E is the elasticity coefficient. We read off the specific strain energy as

$$e(\mathcal{V}) = \frac{EI_0}{2}(v'')^2$$

and the condition of constant such energy, expressed in Theorem 8.1, reads

$$|u''| = c \tag{8.37}$$

for some constant c.

The optimal solution is governed by (8.36) and (8.37) together with equilibrium equations. We find these by looking at the Gateaux derivative of $J(\rho, v)$:

$$J'(\rho, u; \varphi) = EI_0 \int_0^1 \rho u'' \varphi'' \, dx - \int_0^1 q\varphi \, dx = 0,$$

for all φ such that $\varphi(0) = \varphi(1) = 0$. We integrate by parts twice to obtain

$$\left[EI_0 \rho u'' \varphi'\right]_0^1 - \left[(EI_0 \rho u'')' \varphi\right]_0^1 + \int_0^1 \left[(EI_0 \rho u'')'' - q\right] \varphi \, dx = 0,$$

where the second term vanishes since $\varphi(0) = \varphi(1) = 0$. By appropriate choices of φ and by using Lemma 8.1 we find

$$(\rho u'')'' = \frac{q}{EI_0} \quad \text{in } (0, 1), \tag{8.38}$$

$$\rho u'' = 0 \quad \text{at } x = 0 \text{ and } x = 1. \tag{8.39}$$

The boundary condition (8.39) says that the ends of the beam are moment free.

We now solve the above numbered equations to obtain the optimal distribution of the beam width ρ. Equation (8.38) and the boundary conditions (8.39) give

$$\rho u'' = -\frac{q}{2EI_0} x(1 - x).$$

Since the right-hand side of this equation is nonpositive and since ρ is nonnegative, the optimality condition (8.37) gives

$$\rho = Cx(1 - x),$$

for some constant C. The value of this constant is found by substituting into (8.36). We find

$$\rho = \frac{3V}{t} x(1 - x).$$

8.4 Exercises

Exercise 8.1 Find the function u that minimizes

$$J(u) = \int_0^1 \left(1 + (u'')^2\right) dx,$$

subject to the boundary conditions $u(0) = 0$, $u'(0) = 1$, $u(1) = 1$ and $u'(1) = 1$.

Exercise 8.2 As indicated in Sect. 8.1, calculus of variations can be used to show that the shortest path between two points in a plane is a straight line. The problem of finding the shortest path was formulated as

$$\begin{cases} \min\limits_{u} & \int_a^b \sqrt{1 + (u')^2}\, dx \\ \text{s.t.} & \begin{cases} u_a = u(a) \\ u_b = u(b). \end{cases} \end{cases}$$

Find the function that solves this problem and show that it is a straight line.

Exercise 8.3 For the Euler-Bernoulli beams shown in Fig. 8.13, write down the potential energies and derive from them the corresponding ($\mathbb{PDE}$) formulations.

Exercise 8.4 How would the shape of the optimal rod in Sect. 8.3.2 change if the constant load b would be negative?

Exercise 8.5 How would the shape of the optimal rod in Sect. 8.3.2 change if the elasticity coefficient E is a linear function of x?

Exercise 8.6 Calculate the equilibrium displacement distribution corresponding to the optimal shape of the rod in Sect. 8.3.2.

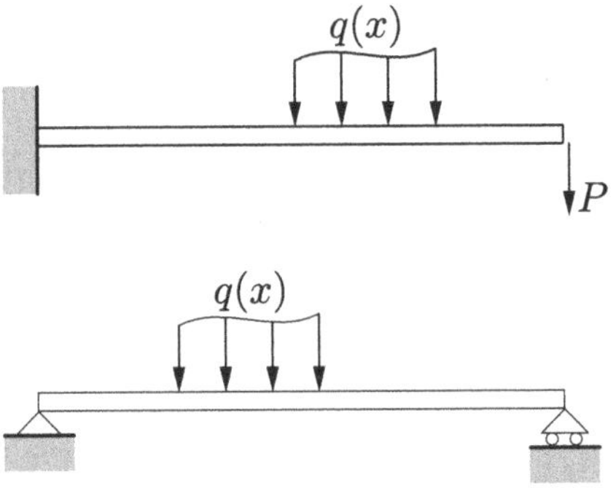

Fig. 8.13 The two different beams of Exercise 8.3

Exercise 8.7 An elastic rod of length L and Young's modulus E, shown in Fig. 8.14, is fixed at $x = 0$ and subject to a force of magnitude $b_0 L$, for a constant b_0, at $x = L$. The rod is also subjected to a body force (per unit length)

$$b(x) = b_0(1 - x/L).$$

Find the cross-sectional area function $\rho(x)$ that maximizes stiffness subject to the constraint

$$\int_0^L \rho\, dx = V$$

on available volume.

Exercise 8.8 Consider an elastic rod of length $2L$ and Young's modulus E that is fixed at both $x = 0$ and $x = 2L$, and which is shown in Fig. 8.15. The rod is subjected to a constant body force (per unit length) b. Find the cross-sectional area function $\rho(x)$ that maximizes stiffness subject to the constraint

$$\int_0^{2L} \rho\, dx = 2V$$

on available volume. The equilibrium conditions that one routinely derives by minimizing potential energy assumes sufficient smoothness of the displacement field

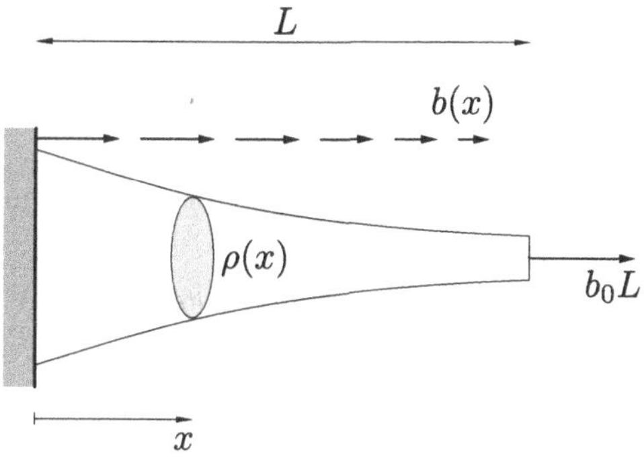

Fig. 8.14 The rod of Exercise 8.7

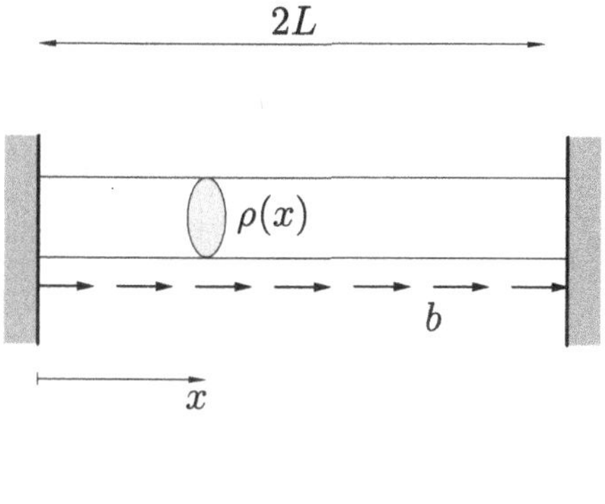

Fig. 8.15 The rod of Exercise 8.8

u and the design ρ. At a point where u and/or ρ is continuous but not differentiable we may use the equilibrium condition

$$(\rho E u')_+ = (\rho E u')_-,$$

where $+$ and $-$ denotes left and right limits.

Exercise 8.9 Verify by means of examples the claim made in the beginning of Sect. 8.3.3 that "in any beam problem the magnitude of the displacement is proportional to the inverse of the area moment of inertia."

Exercise 8.10 Consider the cantilever beam in the upper part of Fig. 8.13. Let $q = 0$ and $P > 0$, and let the beam be of length L and having a rectangular cross section with constant height h and variable width $\rho(x)$. Suppose that the available volume is V and

1. calculate the distribution $\rho(x)$ that gives optimal stiffness.
2. calculate the distribution $\rho(x)$ that gives a constant distribution of maximum normal stress over the length of the bar. Compare with 1.

Exercise 8.11 A beam of length L is freely supported at its both ends (at $x = 0$ and $x = L$), see Fig. 8.16. It is subjected to a constant load intensity (per unit length) $q(x) = q_0 > 0$ and a couple $M_0 > 0$ is applied at its right end (at $x = L$).

Suppose that the beam has a rectangular cross section with a fixed height h and a variable width ρ. Then the total potential energy is given by the functional

$$\mathcal{J}(\rho, v) = \frac{Eh^3}{24}\int_0^L \rho(v'')^2\,dx - \int_0^L qv\,dx + M_0 v'(L),$$

where E is the Young's modulus.

Find the width function $\rho(x)$ that gives maximum stiffness, given the constraint

$$\int_0^L h\rho(x)\,dx = V$$

on available volume.

Exercise 8.12 A shaft of length L is fixed at one end and subjected to a twisting moment M at the other end, see Fig. 8.17. The rotation of a cross section is denoted κ

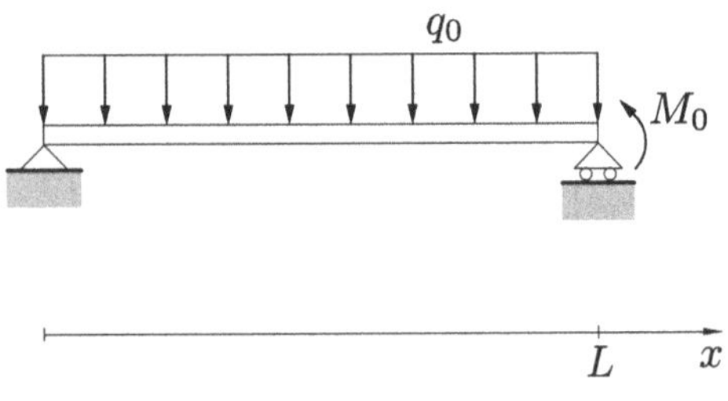

Fig. 8.16 The beam of Exercise 8.11

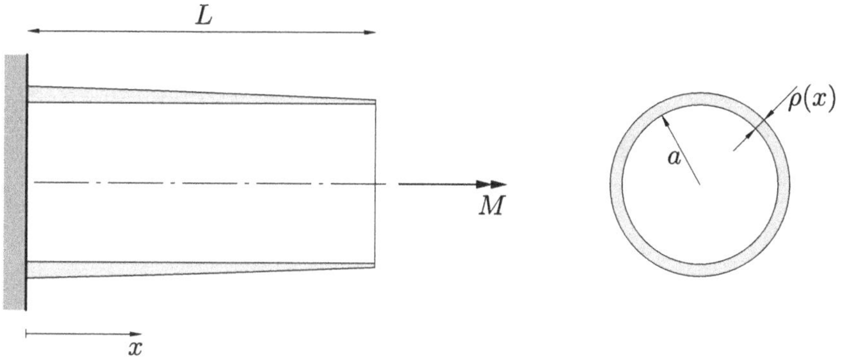

Fig. 8.17 The thin-walled cylinder of Exercise 8.12

and is a function of the coordinate x. The shape of each cross section is a thin-walled cylinder with radius a and material thickness $\rho = \rho(x)$. The total potential energy of such a system is given by the functional

$$\mathcal{J}(\rho, \kappa) = \frac{1}{2} \int_0^L GK(\rho) \left(\frac{d\kappa}{dx} \right)^2 dx - M\kappa(L),$$

where G is the shear modulus and

$$K(\rho) = 2\pi a^3 \rho$$

is the polar moment of inertia.

Find the material thickness distribution ρ that gives maximum stiffness, given the constraint

$$2\pi a \int_0^L \rho \, dx = V$$

on available volume.

Chapter 9
Topology Optimization of Distributed Parameter Systems

This chapter gives a brief introduction to formulations and solution techniques for topology optimization of elastic structures. As a starting point we formulate the problem of optimizing stiffness of a sheet by finding an optimal thickness distribution, which is basically a special case of the general stiffness optimization problem of the previous chapter and which relates closely to the truss problem of Chap. 5. The classical optimality criteria method has shown to be very efficient and is widely used for problems of this type. We show that this method can be seen as a special case of the sequential convex approximation method of Chap. 4. Formulations and solution techniques for topology optimization are next introduced as a modification of the variable thickness sheet problem where penalization is introduced to favor discrete-valued thickness distributions. We discuss the occurrence of ill-posedness of formulations and numerical instabilities, and possible cures of these difficulties based on restriction or relaxation. As a standard reference for structural topology optimization we mention Bendsøe and Sigmund [4].

9.1 The Variable Thickness Sheet Problem

The problem that we formulate here is a stiffness optimization problem of the type previously discussed in the discrete setting of a truss in Chap. 5 and in the continuous setting in Chap. 8. We perform a finite element discretization and show how to solve the problem by an optimality criteria method.

9.1.1 Problem Statement and FE-Discretization

The stiffness optimization problem of this section is the variable thickness sheet problem, i.e., a case of two-dimensional elasticity, as shown in Fig. 8.6, where the thickness is taken as the design function, see Fig. 9.1.

This problem can be seen as a slight generalization of the abstract design problem introduced in Sect. 8.3; the generalization being that we introduce an upper bound $\overline{\rho}$ and a nonzero lower bound $\underline{\rho} > 0$ on the design or thickness function ρ. That is, the set $\mathcal{H}$ is now defined by

$$\begin{cases} \int_{\Omega} \rho \, d\Omega = V \\ \underline{\rho} \leq \rho(x) \leq \overline{\rho} \quad \text{for all } x \in \Omega. \end{cases} \iff \rho \in \mathcal{H}.$$

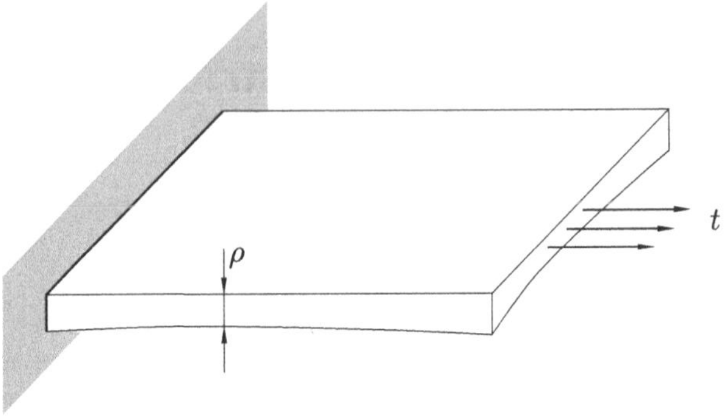

Fig. 9.1 The variable thickness sheet problem where we are seeking for an optimal thickness distribution $\rho(x)$

Thus, we are to solve the following problem:

$$(\mathbb{P}_s^{\text{sheet}}) \quad \begin{cases} \min\limits_{\boldsymbol{u},\rho} \ \ell(\boldsymbol{u}) \\ \text{s.t.} \quad \begin{cases} \boldsymbol{u} \in K \text{ such that } \quad a(\rho, \boldsymbol{u}, \boldsymbol{v}) = \ell(\boldsymbol{v}) \quad \text{for all } \boldsymbol{u} \in K \\ \rho \in \mathcal{H}. \end{cases} \end{cases}$$

Here

$$a(\rho, \boldsymbol{u}, \boldsymbol{v}) = \int_{\Omega} (\nabla \boldsymbol{u})^T \rho \boldsymbol{D} \nabla \boldsymbol{v} \, dA,$$

where $\boldsymbol{D}$ is given in (8.25) for an isotropic material in plane stress; assuming no volume forces, the compliance is given by

$$\ell(\boldsymbol{v}) = \int_{\Gamma_t} \boldsymbol{v}^T \boldsymbol{t} \, ds,$$

and the set of admissible displacements is defined as

$$K = \{\boldsymbol{v} : \Omega \to \mathbb{R}^2 \mid \boldsymbol{v} = \boldsymbol{0} \text{ on } \Gamma_{\mathbf{u}}\}.$$

We introduce a Finite Element (FE) discretization of $(\mathbb{P}_s^{\text{sheet}})$. The resulting problem will then be mathematically equivalent to the truss problem treated in Chap. 5, which may be utilized in several places.

The two-dimensional domain Ω is divided into finite elements Ω_e, $e = 1, \ldots, n$. Using standard FE interpolation for displacement fields and approximating ρ as *constant* in each finite element, a typical element stiffness matrix can be written

$$\boldsymbol{k}_e(x_e) = x_e \boldsymbol{k}_e^0,$$

where x_e is the approximate value of ρ in Ω_e and $\boldsymbol{k}_e^0$ is the element stiffness matrix for unit thickness. Then the global stiffness matrix can be written as

$$\boldsymbol{K}(\boldsymbol{x}) = \sum_{e=1}^{n} x_e \boldsymbol{K}_e^0$$

where $\boldsymbol{x} = [x_1, \ldots, x_n]^T$ is a vector of approximate thicknesses and

$$\boldsymbol{K}_e^0 = \boldsymbol{C}_e^T \boldsymbol{k}_e^0 \boldsymbol{C}_e$$

is a global version of the unit thickness stiffness matrix of element e. Here, $\boldsymbol{C}_e$ is a kinematic matrix that changes local degrees-of-freedoms for global ones, i.e., performs the assembly process. Note that matrices analogous to these where introduced in Chap. 5.

The introduction of a finite element approximation of the displacement field means that the compliance is approximated as

$$\int_{\Gamma_t} \boldsymbol{u}^T \boldsymbol{t}\, ds \approx \boldsymbol{F}^T \boldsymbol{u},$$

where $\boldsymbol{u}$ is the vector of nodal displacements and $\boldsymbol{F}$ is the vector of nodal forces. Note that we use the same notation for both the field of displacements and the vector of nodal displacements. Moreover, the equilibrium problem becomes

$$\boldsymbol{K}(\boldsymbol{x})\boldsymbol{u} = \boldsymbol{F}.$$

Finally, the assumption of constant thickness in each finite element introduces the following approximation of the integral in the volume constraint:

$$\int_{\Omega} \rho\, d\Omega = \sum_{e=1}^{n} \int_{\Omega_e} \rho\, dA \approx \sum_{e=1}^{n} x_e a_e = \boldsymbol{x}^T \boldsymbol{a},$$

where a_e is the area of element Ω_e and $\boldsymbol{a} = [a_1, \ldots, a_n]^T$ is a vector of such areas.

An FE-discretized version of $(\mathbb{P}_s^{\text{sheet}})$ now becomes:

$$(\mathbb{P}_s^{\text{sheet}})^{\text{FE}} \quad \begin{cases} \min\limits_{\boldsymbol{u},\boldsymbol{x}} \ \boldsymbol{F}^T \boldsymbol{u} \\ \text{s.t.} \ \begin{cases} \boldsymbol{K}(\boldsymbol{x})\boldsymbol{u} = \boldsymbol{F} \\ \boldsymbol{x}^T \boldsymbol{a} = V \\ \underline{\rho} \le x_e \le \overline{\rho}, \quad e = 1, \ldots, n. \end{cases} \end{cases}$$

A nested version of this problem is

$$(\mathbb{P}_s^{\text{sheet}})_{\text{nf}}^{\text{FE}} \quad \begin{cases} \min\limits_{\boldsymbol{x}} \ C(\boldsymbol{x}) = \boldsymbol{F}^T \boldsymbol{u}(\boldsymbol{x}) \\ \text{s.t.} \ \begin{cases} \boldsymbol{x}^T \boldsymbol{a} = V \\ \underline{\rho} \le x_e \le \overline{\rho}, \quad e = 1, \ldots, n, \end{cases} \end{cases}$$

where $\boldsymbol{u}(\boldsymbol{x}) = \boldsymbol{K}(\boldsymbol{x})^{-1}\boldsymbol{F}$.

These two problems are mathematically equivalent to the two problems introduced in relation to trusses in Chap. 5. They could therefore be solved using MMA as outlined there.

Example 9.1 We wish to determine the thickness distribution that minimizes the compliance of the sheet in Example 7.4. The mesh is depicted in Fig. 7.27 and consists of 1800 elements. Since we are dealing with a sizing optimization problem, this mesh will be fixed throughout the iterations. The weight of the sheet is fixed to 180 units, and the density is 1. The thickness is taken to be constant over each finite element, and has to be greater than 0.05, but smaller than 5. These thicknesses are the design variables and, thus, there are 1800 such variables in this problem. The problem is solved using the MMA method described in Chap. 5. Figure 9.2 shows how the thickness distribution changes during the MMA iterations. After 40 iterations, the solution barely changes at all. Figure 9.3 illustrates how the compliance and the weight change during the optimization.

9.1.2 The Optimality Criteria (OC) Method

A classical approach to the numerical solution of a discretized structural optimization problem of the type given in $(\mathbb{P}_s^{\text{sheet}})_{\text{nf}}^{\text{FE}}$, is the optimality criteria (OC) method. This method is historically older than the method of explicit convex approximations introduced in Chap. 4 and the two classes of methods are frequently conceived as competing alternative approaches. However, in this section we will show that, at least for the special problem $(\mathbb{P}_s^{\text{sheet}})$, the OC method can be seen as a special case of the explicit convex approximation method. The reason for introducing the OC method is that it has turned out to be very efficient for solving the topology optimization problems discussed in later sections of this chapter.

Note that $(\mathbb{P}_s^{\text{sheet}})_{\text{nf}}^{\text{FE}}$ is very similar to the problem that was studied for trusses in Chap. 5, so from (5.24), we therefore have that

$$\frac{\partial C(\boldsymbol{x})}{\partial x_e} = -\boldsymbol{u}_e(\boldsymbol{x})^T \boldsymbol{k}_e^0 \boldsymbol{u}_e(\boldsymbol{x}) = -(\boldsymbol{u}_e^k)^T \boldsymbol{k}_e^0 \boldsymbol{u}_e^k \quad \text{at } \boldsymbol{x} = \boldsymbol{x}^k, \tag{9.1}$$

where $\boldsymbol{u}_e^k = \boldsymbol{u}_e(\boldsymbol{x}^k) = \boldsymbol{C}_e \boldsymbol{u}(\boldsymbol{x}^k)$ and $\boldsymbol{u}(\boldsymbol{x}^k) = \boldsymbol{K}(\boldsymbol{x}^k)^{-1}\boldsymbol{F}$.

The idea that reveals the connection between OC and the method of explicit convex approximations is a linearization of $C(\boldsymbol{x})$ in the intervening variable

$$y_e = x_e^{-\alpha},$$

where α is any number greater than zero. Since, due to positive definiteness of stiffness matrices, the derivative (9.1) is always nonpositive and, therefore, $\alpha = 1$ corresponds to the CONLIN linearization in Sect. 4.4. Using this intervening variable

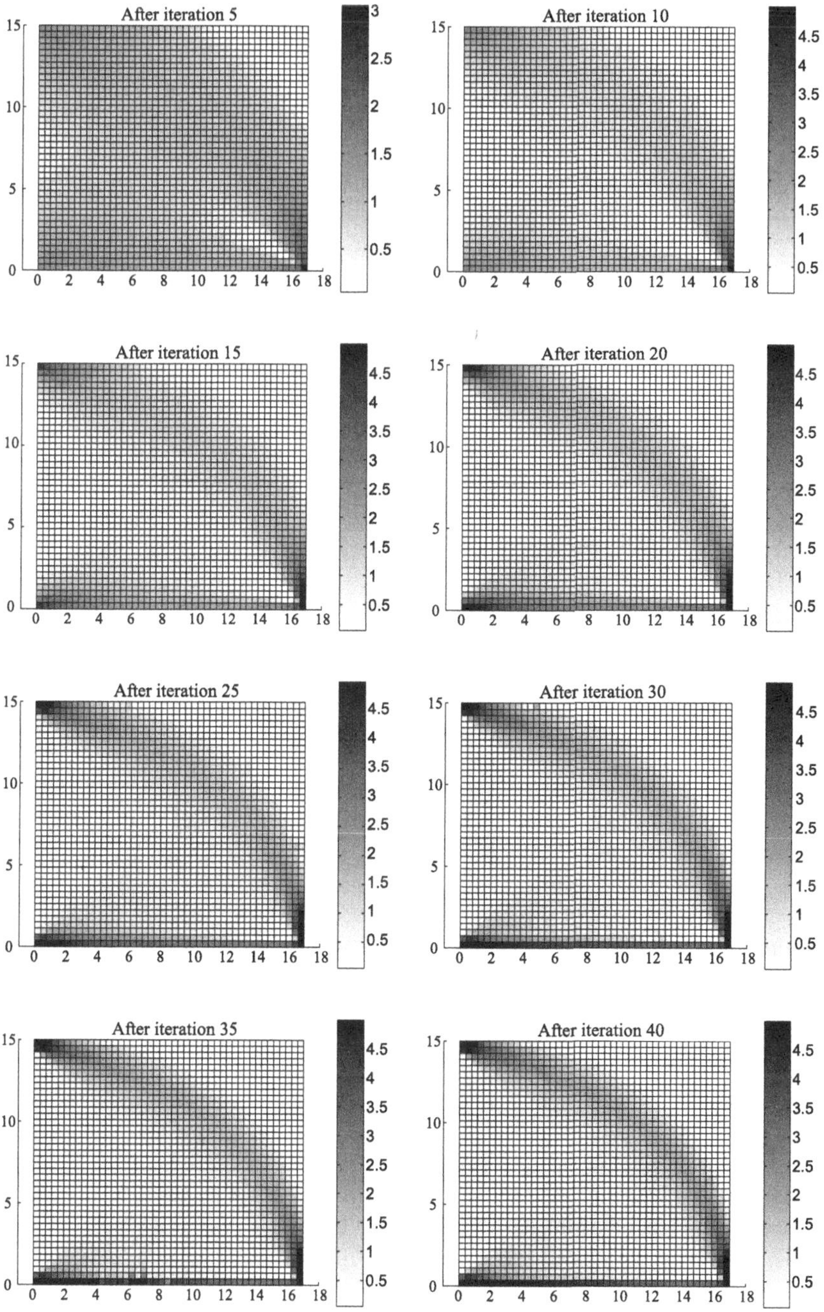

Fig. 9.2 The thickness distribution of the sheet at various iterations

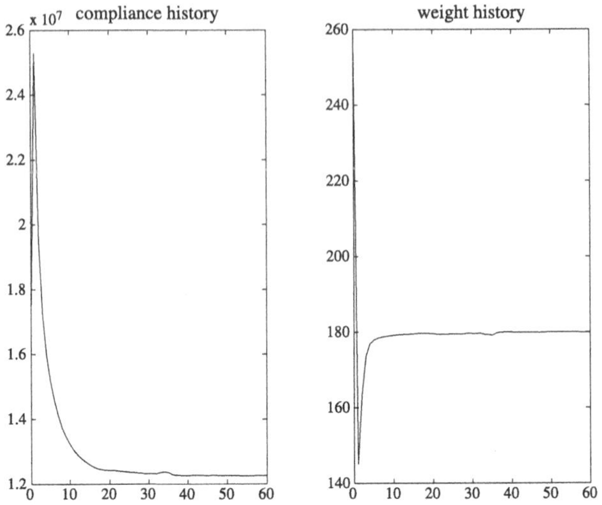

Fig. 9.3 The history of the compliance and the weight

we then get

$$C(\boldsymbol{x}) \approx C(\boldsymbol{x}^k) + \sum_{e=1}^{n} \left.\frac{\partial C}{\partial y_e}\right|_{\boldsymbol{x}=\boldsymbol{x}^k} (y_e - y_e^k), \tag{9.2}$$

where

$$\frac{\partial C}{\partial y_e} = \frac{\partial C}{\partial x_e}\frac{\partial x_e}{\partial y_e} = \frac{\partial C}{\partial x_e}\frac{1}{\frac{dx_e^{-\alpha}}{dx_e}} = -\frac{1}{\alpha x_e^{-\alpha-1}}\frac{\partial C}{\partial x_e} = -\frac{x_e^{1+\alpha}}{\alpha}\frac{\partial C}{\partial x_e}. \tag{9.3}$$

Inserting (9.1) and (9.3) in (9.2) gives

$$C(\boldsymbol{x}) \approx \text{const.} + \sum_{e=1}^{n} b_e^k x_e^{-\alpha},$$

where

$$b_e^k = \frac{1}{a}\left((\boldsymbol{u}_e^k)^T \boldsymbol{k}_e^0 \boldsymbol{u}_e^k\right)(x_e^k)^{1+a}. \tag{9.4}$$

We can now formulate an approximate subproblem:

$$(\mathbb{P}_s^{\text{sheet}})_k^{\text{FE}} \quad \begin{cases} \min\limits_{\boldsymbol{x}} \ \sum\limits_{e=1}^{n} b_e^k x_e^{-\alpha} \\ \text{s.t.} \ \begin{cases} \boldsymbol{x}^T \boldsymbol{a} = V \\ \underline{\rho} \le x_e \le \overline{\rho}, \quad e = 1, \ldots, n. \end{cases} \end{cases}$$

This problem is convex and can be treated by Lagrangian duality. The Lagrangian function is

$$\mathcal{L}(\boldsymbol{x}, \lambda) = \sum_{e=1}^{n} b_e^k x_e^{-\alpha} + \lambda \left(\boldsymbol{x}^T \boldsymbol{a} - V \right).$$

The dual objective function is

$$\varphi(\lambda) = \min_{\underline{\rho} \le x_e \le \overline{\rho}} \mathcal{L}(\boldsymbol{x}, \lambda) = \sum_{e=1}^{n} \min_{\underline{\rho} \le x_e \le \overline{\rho}} \left[b_e^k x_e^{-\alpha} + \lambda a_e x_e \right] - \lambda V.$$

Clearly, this function has a separable structure and $\varphi(\lambda)$ can be evaluated by finding separate minima for n functions

$$\varphi_e(x_e, \lambda) = b_e^k x_e^{-\alpha} + \lambda a_e x_e.$$

Assuming that the minimum of $\varphi_e(x_e, \lambda)$ is taken inside the interval $\underline{\rho} \le x_e \le \overline{\rho}$, we can find it by seeking a stationary point:

$$\frac{\partial \varphi_e(x_e, \lambda)}{\partial x_e} = -\alpha b_e^k x_e^{(-\alpha-1)} + \lambda a_e = 0,$$

from which we find

$$x_e = \left(\frac{\alpha b_e^k}{\lambda a_e} \right)^{\frac{1}{1+\alpha}}.$$

By studying this value we can determine whether it was a correct assumption that the minimum was inside the interval and the primal-dual relation becomes

$$x_e(\lambda) = \begin{cases} \underline{\rho} & \text{if } \left(\frac{\alpha b_e^k}{\lambda a_e} \right)^{\frac{1}{1+\alpha}} < \underline{\rho} \\ \left(\frac{\alpha b_e^k}{\lambda a_e} \right)^{\frac{1}{1+\alpha}} & \text{if } \underline{\rho} \le \left(\frac{\alpha b_e^k}{\lambda a_e} \right)^{\frac{1}{1+\alpha}} \le \overline{\rho} \\ \overline{\rho} & \text{if } \left(\frac{\alpha b_e^k}{\lambda a_e} \right)^{\frac{1}{1+\alpha}} > \overline{\rho}. \end{cases} \tag{9.5}$$

This function is shown graphically in Fig. 9.4, where it is also indicated in what intervals particular derivatives must be zero.

We can now insert $x_e(\lambda)$ into the Lagrangian function to obtain the dual function $\varphi(\lambda)$ explicitly. The dual problem becomes

$$\max_{\lambda \in \mathbb{R}} \varphi(\lambda),$$

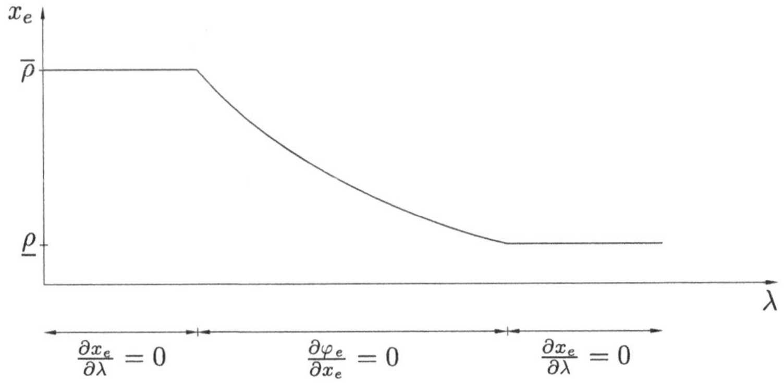

Fig. 9.4 The primal-dual function $x_e(\lambda)$

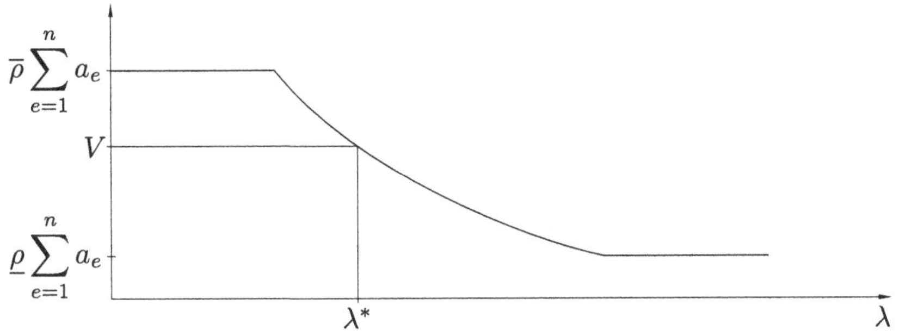

Fig. 9.5 Graphic illustration of the root λ^* that solves (9.6)

where

$$\varphi(\lambda) = \mathcal{L}(\boldsymbol{x}(\lambda), \lambda) = \sum_{e=1}^{n} \varphi_e(x_e(\lambda), \lambda) - \lambda V.$$

We solve the dual problem by seeking a stationary point of $\varphi(\lambda)$. We then use the properties of zero partial derivatives shown in Fig. 9.4:

$$\frac{\partial \varphi(\lambda)}{\partial \lambda} = \sum_{e=1}^{n} \left(\frac{\partial \varphi_e}{\partial x_e} \frac{\partial x_e}{\partial \lambda} + \frac{\partial \varphi_e}{\partial \lambda} \right) - V = \sum_{e=1}^{n} a_e x_e(\lambda) - V = 0. \tag{9.6}$$

Thus, the dual has a stationary value when the volume constraint is satisfied. Equation (9.6) is easily solved since the function that should be zero is monotonically increasing as illustrated in Fig. 9.5. We can find a unique λ^* that solves (9.6) by, e.g., interval reduction.

The solution of $(\mathbb{P}_s^{\text{sheet}})_k^{\text{FE}}$ is now given by inserting λ^* into (9.5) and these values of the thicknesses are taken as the next iterate of the method, i.e.,

$$x_e^{k+1} = x_e(\lambda^*).$$

If we consider only the middle alternative in (9.5) we find by using (9.4) that

$$x_e^{k+1} = \left(\frac{(\boldsymbol{u}_e^k)^T \boldsymbol{k}_e^0 \boldsymbol{u}_e^k}{\lambda a_e} \right)^{\frac{1}{1+\alpha}} x_e^k.$$

We see that the number a only appears in the combination $1/(1+a)$ and we therefore introduce the notation

$$\eta = \frac{1}{1+\alpha},$$

which is called a damping factor. The reason for this terminology will appear shortly.

One step in the iteration method can now be described as follows:

Given a design (thickness) $\boldsymbol{x}^k$, solve equilibrium, i.e., find $\boldsymbol{u}^k$ such that

$$\boldsymbol{K}(\boldsymbol{x}^k)\boldsymbol{u}^k = \boldsymbol{F}.$$

A new design iterate is then given by

$$x_e^{k+1} = \min\left\{ \max\left[x_e^k \left(\frac{(\boldsymbol{u}_e^k)^T \boldsymbol{k}_e^0 \boldsymbol{u}_e^k}{\lambda a_e} \right)^{\eta}, \underline{\rho} \right], \overline{\rho} \right\},$$

where λ is determined by

$$\sum_{e=1}^{n} a_e x_e^{k+1}(\lambda) - V = 0.$$

Traditionally this is called an optimality criteria (OC) method. Such methods were initially formulated directly on the following intuitive grounds: If the constraint $\underline{\rho} \leq x_e \leq \overline{\rho}$ is inactive, the method has converged if

$$\left(\frac{(\boldsymbol{u}_e^k)^T \boldsymbol{k}_e^0 \boldsymbol{u}_e^k}{\lambda a_e} \right) = 1.$$

This means that

$$\frac{(\boldsymbol{u}_e^k)^T \boldsymbol{k}_e^0 \boldsymbol{u}_e^k}{a_e} = \lambda = \text{constant} \quad \text{for all } e = 1, \ldots, n.$$

The quotient represents twice the strain energy per volume, or specific strain energy, which is shown to be constant for every finite element at convergence, so the iteration of the method attempts to modify thicknesses so as to approach such a state.

Elements with high strain energy is expected to be low on stiffness, so we make these elements thicker. When η is less than unity this modification is damped and, hence, the term damping factor for η.

9.2 Penalization of Intermediate Thickness Values

The variable thickness sheet problem of the previous section is a sizing optimization problem if the lower thickness bound is nonzero. On the other hand, if this thickness is zero the problem also has a topology optimization character since regions where $\rho = 0$ can be interpreted as new holes in Ω. Taking this observation one step further we may think of a pure topology optimization problem where the sheet thickness can take only the lower bound $\underline{\rho} = 0$ and the upper bound $\overline{\rho}$. From a practical point of view this can concern design of a stamping operation where we are to punch out optimal holes in a sheet of thickness $\overline{\rho}$. Now, unfortunately, there are a number of difficulties associated with this type of discrete optimization problem. Firstly, having $\underline{\rho}$ equal zero produces finite elements with zero stiffness giving state problems that are singular and not uniquely solvable, as required by the nested formulation. Therefore, a standard procedure is to let $\underline{\rho} = \varepsilon$, where ε is a small positive value, but still interpreting regions where the optimal $\rho = \varepsilon$ as holes. Secondly, optimization algorithms for problems where variables can take only discrete values are not very efficient for the large problems that we envisage here. Therefore, an approach where intermediate values of the design variables are allowed but penalized should be the more efficient one, and is described in the following. When presenting this approach, we use $\overline{\rho} = 1$. The reason is that the ideas are then directly applicable for three-dimensional elasticity where ρ cannot represent a thickness, but can be thought of as a variable which signals no material ($\rho = \varepsilon \approx 0$) or material ($\rho = 1$).

9.2.1 *Solid Isotropic Material with Penalization (SIMP)*

Solid isotropic material with penalization (SIMP) means that intermediate designs are penalized by using the following constitutive matrix in Hooke's law:

$$\boldsymbol{D} = \frac{\rho^q E}{1-\nu^2}\begin{bmatrix} 1 & \nu & 0 \\ \nu & 1 & 0 \\ 0 & 0 & \dfrac{1-\nu}{2} \end{bmatrix}, \tag{9.7}$$

for a constant q. Thus, the "effective" Young's modulus is $\rho^q E$. Comparing to the variable thickness sheet problem, the SIMP method is obtained by changing ρ for ρ^q and setting $\underline{\rho} = \varepsilon \approx 0$ and $\overline{\rho} = 1$.

The "effective" Young's modulus as a function of ρ for different values of q is shown in Fig. 9.6. We see from this figure that if $q > 1$, then for ρ between 0 and 1 the obtained stiffness is disproportionately low and therefore such values will

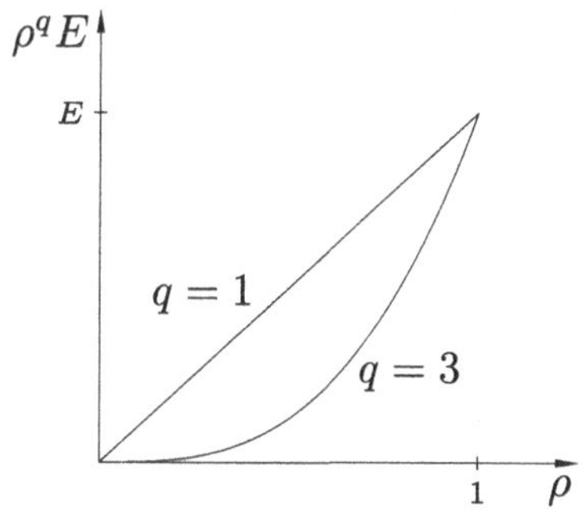

Fig. 9.6 The "effective" Young's modulus as a function of ρ for different values of q

be avoided in the optimal solution: they will not represent an economical use of material. Thus, we may expect an optimal solutions based on the constitutive matrix (9.7), to consists of large regions where $\rho = \varepsilon \approx 0$, i.e., "holes," and large regions where $\rho = 1$, i.e., regions where Young's modulus is E.

SIMP can be used also for a three-dimensional integration domain Ω. We can then no longer interpret ρ as a thickness. However, we can think of it as a sort of generalized density, with the interpretation that $\rho = \varepsilon \approx 0$ represents holes and $\rho = 1$ represents solid regions with a prescribed Young's modulus E. The physical dimension of ρ is then no longer a length, rather it is dimensionless.

We like to see how the OC method for the sheet problem, given in the previous section, needs to be changed to work for the SIMP method. With the constitutive matrix (9.7), the global stiffness matrix becomes

$$\boldsymbol{K}(\boldsymbol{x}) = \sum_{e=1}^{n} x_e^q \boldsymbol{K}_e^0,$$

and its sensitivity derivative is

$$\frac{\partial \boldsymbol{K}(\boldsymbol{x})}{\partial x_e} = q x_e^{q-1} \boldsymbol{K}_e^0.$$

This means that (9.1) should be replaced by

$$\frac{\partial C(\boldsymbol{x})}{\partial x_e} = -\boldsymbol{u}_e(\boldsymbol{x})^T \left\{ q x_e^{q-1} \boldsymbol{k}_e^0 \right\} \boldsymbol{u}_e(\boldsymbol{x}).$$

Hence, the only modification of the OC method on page 187 that we need to do to get the SIMP OC method is to

$$\text{replace } (\boldsymbol{u}_e^k)^T \boldsymbol{k}_e^0 \boldsymbol{u}_e^k \text{ by } (\boldsymbol{u}_e^k)^T \left\{ q (x_e^k)^{q-1} \boldsymbol{k}_e^0 \right\} \boldsymbol{u}_e^k.$$

9.2.2 Other Penalizations

In the SIMP method just described, we replaced the ρ of the variable thickness sheet problem by $\eta(\rho) = \rho^q$ in the constitutive equation. There are other functions that may be used instead of the SIMP scheme to penalize intermediate values. For instance

$$\eta(\rho) = \frac{\rho}{1 + (q-1)(1-\rho)},$$

where q should be greater that one and is a control of the level of penalization.

Another way to penalize intermediate values is to keep the original $\eta(\rho) = \rho$ in the constitutive equation, but adding a penalty term to the objective function. A penalty function that could be used for this is

$$P(\rho) = \int_{\Omega} (\rho - \underline{\rho})(\overline{\rho} - \rho)\, d\Omega, \tag{9.8}$$

which is always positive. This function is multiplied by some large constant q and added to the objective function. In the minimization process it should tend to small values, which means that ρ tends towards $\underline{\rho}$ or $\overline{\rho}$.

9.3 Well-Posedness and Potential Numerical Problems

9.3.1 The Archetype Problem and an Analogy

In the previous section we penalized intermediate values in order to achieve a "black and white" design. Our goal was to get an approximate solution of the following problem:

$$(\mathbb{P}_a) \quad \begin{cases} \min\limits_{\boldsymbol{u},\rho} \ \ell(\boldsymbol{u}) \\ \text{s.t.} \quad \begin{cases} \boldsymbol{u} \in K \text{ such that} \quad a(\rho, \boldsymbol{u}, \boldsymbol{v}) = \ell(\boldsymbol{v}) \quad \text{for all } \boldsymbol{u} \in K \\ \displaystyle\int_{\Omega} \rho\, d\Omega = V \\ \text{point values of } \rho \text{ belongs to } \{0, 1\}. \end{cases} \end{cases}$$

This is called the *archetype* problem and the linear and bilinear forms, ℓ and a, could be related to both two- and three-dimensional elasticity with the elasticity modulus ρE. As already indicated, the constraint that point values of ρ should belong to the integer set $\{0, 1\}$ makes this an integer programming problem that is not easy to handle. Therefore, we suggested using penalization. Now, regardless of whether we are to treat the original integer problem or its penalization, it is reasonable to ask if the problem is at all solvable, i.e., is it well-posed in the sense that there actually exists a solution? The answer to this question, perhaps at first surprising, is that for

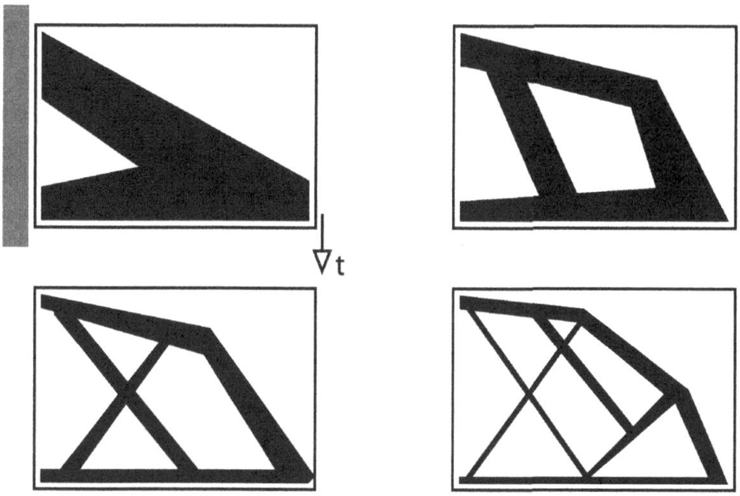

Fig. 9.7 Introducing more and thinner bars gives a better objective function value, but there is no end to this process if the archetype problem or its penalization is treated. Picture by Borrvall

most specific choices of boundary conditions *there exists no solution of* $(\mathbb{P}_a)$! An intuitive explanation of this fact follows below.

In the upper left picture in Fig. 9.7 a design essentially consisting of two bars is shown. If we replace these bars by more but thinner bars, the stiffness can be made larger. As indicated in Fig. 9.7, this process goes on indefinitely. Thus, by adding more and more bars we get a stiffer and stiffer structure with thinner and thinner bars. The situation can be compared to difficulties related to

$$\text{minimizing } f(x) = 1/x \quad \text{where } x \in \mathcal{H} = \{x \in \mathbb{R} \mid x \geq 1\}. \tag{9.9}$$

Clearly this problem has no solution since given a candidate solution $x \in \mathcal{H}$, the objective can be made better by simply taking $x_{\text{new}} = x + 1$, and, as in the archetype problem, this process goes on for ever.

As a remark, note that the variable thickness sheet problem is not included in what is just said. Problem $(\mathbb{P}_s^{\text{sheet}})$ possesses a solution, which is shown in Petersson [25], but when the SIMP modification is introduced to get "black and white" designs, this property disappears with $q > 1$.

9.3.2 Numerical Instabilities

If one disregards the theoretical difficulty of nonexistence, indicated in the previous subsection, and goes on to perform an FE discretization of the archetype problem or its penalization (this is what is done in Sect. 9.2.1, where the SIMP method is described) then several numerical difficulties may occur. These are described below. For a more thorough discussion we refer to Sigmund and Petersson [32] and Borrvall [6].

At first it may seem as if one gets away with disregarding nonexistence, since the problem that is produced by introducing the FE-approximation generally shows existence of solutions. However, if one is not satisfied with the resolution of a design picture and therefore refines the mesh and performs the optimization again, then the new numerical result will generally not be an improved picture of the same design. Rather, the design that is produced will be different for each new mesh: new holes will appear and so forth. This is called *mesh-dependency*. Typically, one will get more and thinner structural parts (where $\rho = 1$) as the mesh becomes finer, since nothing restricts this. A consequence is that such parts will tend to be one or two elements wide and thereby result in a structure that is artificially stiff.

Even if one treats a problem for which there exists a solution, like the variable thickness sheet problem, a sequence of FE optimization problems, discretized by finer and finer meshes, may not converge. This happens if the discretizations of the two fields, the design ρ and the displacement $\boldsymbol{u}$, are not carefully chosen. Typically a so-called *checkerboard pattern* appears in such circumstances, where the design function ρ is alternating between 0 and 1, i.e., solid and void, see Fig. 9.8. Moreover, for a problem which has no optimal solution, checkerboards will certainly appear, since then there is even nothing to converge to. In this case there is no correct FE discretization whatsoever. For the variable thickness sheet problem, it was shown in Petersson [26] that an FE discretization with nine node Lagrangian elements for $\boldsymbol{u}$ and elementwise constant approximations for ρ, leads to solutions without checkerboards.

A final difficulty for which there seems to be no obvious cure is nonconvexity. The variable thickness sheet problem is a convex problem; convexity is a very desirable property since every local minimum is then also global, and the global minimum is what we are looking for. Unfortunately, penalties that help in producing "black and white" designs will give nonconvex problems: when $q > 1$ in the SIMP method the problem becomes nonconvex. For such problems it is possible that the algorithm terminates to completely different local optima for different starting points. Presently there are no methods that are guaranteed to converge to global minima, at least not for the large size problems that are typical for topology optimization. It seems as if engineering intuition has to be the guide when a design is accepted. The heuristic methods that may be used are: (1) to run the problem several times with different starting points; (2) solve several problems where the value of q is gradually raised from the value 1, which gives a convex problem, to higher values, which give "black and white" designs.

Figure 9.8 shows the same problem solved three times in different ways by the SIMP method and illustrates some of the difficulties discussed in this section. For instance, all three solutions show checkerboards in different regions of the domain. The upper left solution is produced by raising q in three steps, $q = 1, 2, 3$. The upper right solution is obtained by solving directly for $q = 3$. Finally, the lower picture is obtained by using 4 times as many elements as in the first two pictures. Slightly different topologies are indicated in the three pictures showing a mesh-dependency as well as a dependency that could be due to nonconvexity.

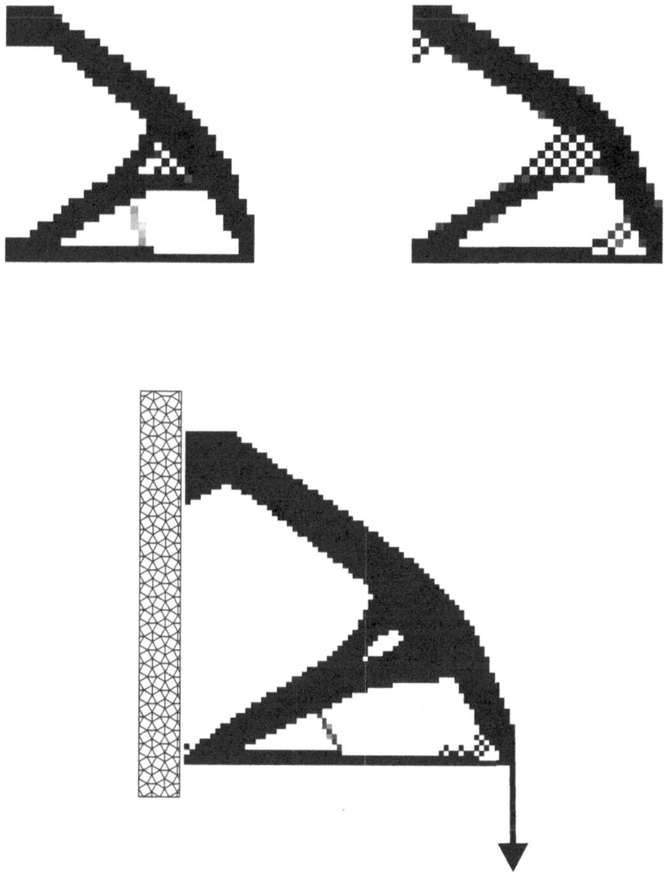

Fig. 9.8 Three pictures produced by use of the SIMP method for an ill-posed problem. The *upper two pictures* are produced using the same FE mesh, but with a continuation procedure for the penalty exponent q in the *left-hand case* but not in the *right-hand case*. The *lower picture* is produced using 4 times the number of elements of the *upper two pictures*. These solutions show mesh-dependency and checkerboards. The solutions were obtained by Torstenfelt

Table 9.1 is a slightly simplified version of the table in Sigmund and Petersson [32] and summarizes what has been discussed above, as well as indicates the cure by restriction to be discussed in the next section.

9.4 Restriction of the Archetype Problem

Nonexistence of solutions to $(\mathbb{P}_a)$ or its penalization occurs since nothing bounds the number of holes or the smallest thickness of structural parts. In Sect. 9.3.1 we considered a simple problem, defined by (9.9), which in analogy with $(\mathbb{P}_a)$ had no solution since the set $\mathcal{H}$ was unbounded to the right. Now, say that one cannot "man-

Table 9.1 Summary of numerical instabilities. An optimal solution to the continuum problem is denoted ρ^* and an optimal design obtained when solving FE-discretized problems are denoted ρ_h^*, where h is a mesh-size

Numerical problems	Mathematical reason	Possible cure
Algorithm terminates at nonglobal minimum.	Nonconvex problem.	Increase q gradually. Try several starting designs.
Checkerboard pattern.	Nonconvergence of ρ_h^* to ρ^* as $h \to 0$.	Increase number of nodes in displacement FE. Use restriction.
Mesh-dependency.	Nonexistence of ρ^*.	Use restriction.

ufacture" x larger than some maximum value c. Then, one would like to replace $\mathcal{H} = \{x \in \mathbb{R} \mid x \geq 1\}$ by $\mathcal{H} = \{x \in \mathbb{R} \mid 1 \leq x \leq c\}$, and with such a set the problem becomes solvable, i.e., well-posed. We call this a *restriction* of the admissible set. Similarly to this we can introduce restrictions on the set of admissible designs in $(\mathbb{P}_a)$ or its penalization and thereby produce well-posed continuum problems that behave well when FE discretized. This section describes some different ways of doing this. The presentation is largely descriptive and we refer to the referenced papers for details.

9.4.1 Bounds on the Design Gradient

A way to measure oscillations of a function is through a norm of its gradient. It has been showed, Borrvall [6], that the family of L^p-norms can be used to reduce the increasingly oscillating designs associated with the archetype problem and, thus, to produce a well-posed problem. That is, the following constraint is added to the SIMP penalized version of $(\mathbb{P}_a)$:

$$\left[\int_{\Omega} |\nabla \rho|^p \, d\Omega\right]^{1/p} \leq C_p, \tag{9.10}$$

where ∇ denotes the gradient, p is any integer such that $1 \leq p \leq \infty$ and $C_p > 0$ is some constant. Instead of adding the L^p-norm as a constraint to $(\mathbb{P}_a)$ it is also possible to multiply it by a penalty parameter and add the result to the objective function. That is, the term

$$c_p \left[\int_{\Omega} |\nabla \rho|^p \, d\Omega\right]^{1/p} \tag{9.11}$$

is added to the objective function for some large constant $c_p > 0$. In fact, this approach has shown to be the more numerically efficient for $p < \infty$.

For the extreme case $p = \infty$, (9.10) reduces to a pointwise bound on the gradient, i.e.,

$$|\nabla \rho| \leq C_\infty.$$

This is an approach used in Petersson and Sigmund [27], where it was shown to give mesh independent FE solutions. It has the advantage of being a local constraint on the width of structural members. In fact, it is shown in Borrvall [6] that the minimum width is approximately $2/C_\infty$. However, the disadvantage is that a large number of constraints have to be handled: instead of one global constraint, as for $p < \infty$, we have a large number of local constraints. Also, designs tend not to be sharply "black and white."

The other extreme case $p = 1$ reduces to a perimeter constraint, i.e., in two dimensions it is a constraint on the length of the boundary between material and holes. This constraint gives nice "black and white" designs with sharp boundaries. However, it is rather unstable and sensitive to local optima and is not recommended in practice.

Considering intermediate values, $1 < p < \infty$, Borrvall [6] recommends $p = 2$ as being easiest to implement and numerically stable.

To study one effect of the choice of the constant C_p we consider a minimization problem with objective function $f(\rho)$. Assume that we solve the problem twice, altering only the value of C_p from C_p^1 to C_p^2, where $C_p^1 < C_p^2$. Then, if the admissible sets of the two versions of the problems are denoted $\mathcal{H}^1$ and $\mathcal{H}^2$, we have that $\mathcal{H}^1 \subset \mathcal{H}^2$, and, therefore, $f(\rho_2^*) \leq f(\rho_1^*)$, where ρ_1^* and ρ_2^* are optimal solutions of the two problems. Thus, in a stiffness optimization problem the structure becomes stiffer as the value of C_p is raised. When the stiffness is raised it is also likely that the structure will consist of thinner and more structural members. These conclusions hold, of course, only if the error due to the FE modeling is disregarded, or if we use the same FE mesh in both cases. For large values of C_p, members may tend to become one or two elements thick, resulting in artificially stiff structures due to poor FE-modeling.

9.4.2 Filters

In image processing, high frequency components are reduced by low pass filtering. Something similar can be done to reduce oscillations in designs associated with the archetype problem. The effect of a filter operator S_R is shown in Fig. 9.9. The filter operator depends on a filter radius R which determines how much of the oscillations are kept. It can be explicitly defined by a convolution, so that the filtered design at a point $x \in \Omega$ is defined by

$$S_R(\rho)(x) = \int_\Omega \rho(y)\phi(x, y)\, d\Omega,$$

where integration is with respect to the variable $y \in \Omega$ and the so-called filter $\phi(x, y)$ could be

$$\phi(x, y) = \frac{3}{\pi R^2} \max\left(0, 1 - \frac{|x - y|}{R}\right), \tag{9.12}$$

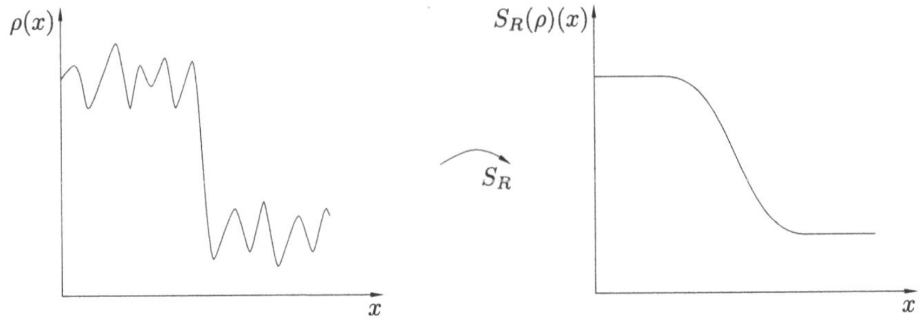

Fig. 9.9 The filter operator removes oscillations. The filter radius R controls how local an oscillation should be to be removed

Fig. 9.10 The filter function (9.12)

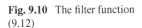

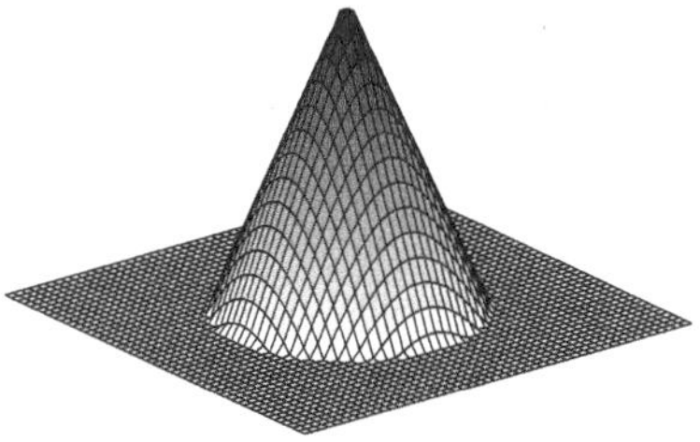

which is shown in Fig. 9.10.

There are at least three different ways of including the filter operator in the archetype problem. A simple but efficient approach was suggested by Sigmund [30, 31]. It consists of filtering the sensitivities in an OC or a convex approximation approach. In contrast to the two other filter methods described below there is no mathematical proof that this method will actually result in mesh independent designs, but all numerical experience points in this direction.

Bruns and Torterelli [10] introduced filtering by changing the SIMP interpolation function $\eta(\rho)$ for $\eta(S_R(\rho))$. Then the minimum width of a structural member becomes approximately $2\pi R/3$. This method has the disadvantage of introducing an ambiguity as to which density to plot: the optimization is done for the unfiltered density but the structural response is calculated for the filtered one.

Borrvall and Petersson [7] introduced filters by means of the penalty function (9.8). They kept the linear dependence on ρ in the equilibrium equation and added a penalty term to the objective function. However, the penalty function $P(\rho)$ is not regular enough to result in a well-posed problem: the same difficulties as with the original SIMP method occur. Therefore, a *regularized penalty* in the form $P(S_R(\rho))$ was used, which does result in a well-posed problem and a well behaved FE discretization. Since contributions to the penalty function comes from the region between "black and white" in a solution, this method is very similar to the perimeter constraint of the previous subsection, but seems to be more numerically stable. Although regularized penalty works well for the stiffness optimization problem, a disadvantage, compared to using bounds on the design variable, is that it is not clear that it can be extended to other choices of objective function.

Fig. 9.11 Convergence study for the regularized penalty method. The loading and boundary conditions were given in Fig. 1.5. The solutions were obtained by Borrvall and Petersson

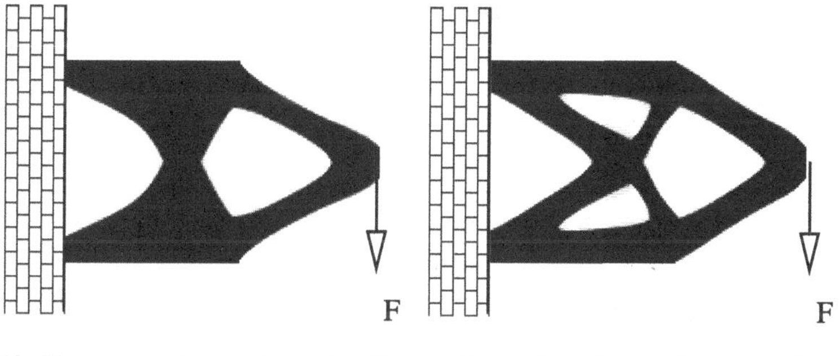

Fig. 9.12 The same problem solved with different filter radius. The solutions were obtained by Borrvall and Petersson

Figure 9.11 shows four different solutions for a so-called MBB beam, using regularized penalty. These solutions where presented in Borrvall and Petersson [7]. The upper solution is for 2 400 elements, the next one uses 9 600 elements and the two lower solutions are for 38 400 elements. The bottom solution uses a larger penalty parameter than in the other three cases.

The solutions in Fig. 9.12 are also from Borrvall and Petersson [7] and show the dependency on the filter radius R. In both solutions 16 000 elements are used.

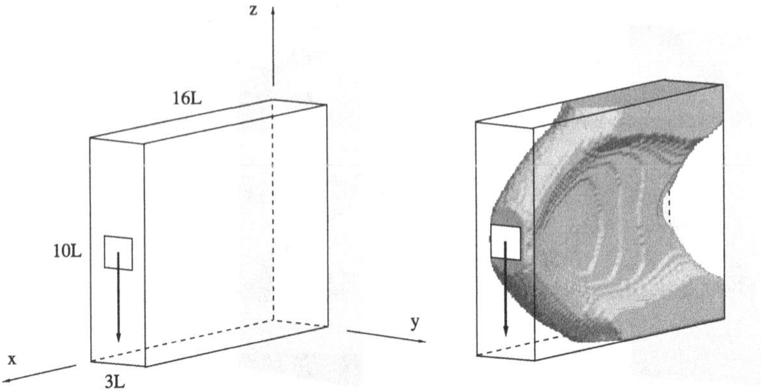

Fig. 9.13 A three-dimensional cantilever beam. The filter radius is $R = 0.5L$, and the available volume is 50% of the box in the *left-hand picture*. The solution was obtained by Borrvall and Petersson

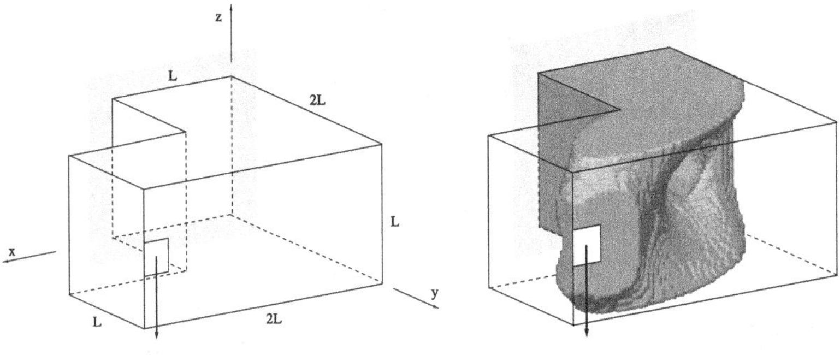

Fig. 9.14 A three-dimensional crank structure. The filter radius is $R = 0.1L$, and the available volume is 50% of the box in the *left-hand picture*. The solution was obtained by Borrvall and Petersson

Finally in this section we give some three-dimensional results from Borrvall and Petersson [8], again using regularized penalty. Figure 9.13 shows a cantilever beam using 245 760 elements. Figure 9.14 shows a crank type structure that was obtained using 192 000 elements. In Fig. 9.15 the same solution is shown from different views. Note the I-beam type structure of the outer part of the structure, while close to the wall the optimum shape tends to have a circular hole.

9.5 Relaxation of the Archetype Problem

Consider the simple problem (9.9), used for analogy in Sect. 9.3.1. This problem does not have a solution, but it was noted in the introduction of the previous section that it could be made solvable by restricting the admissible set. Another way of

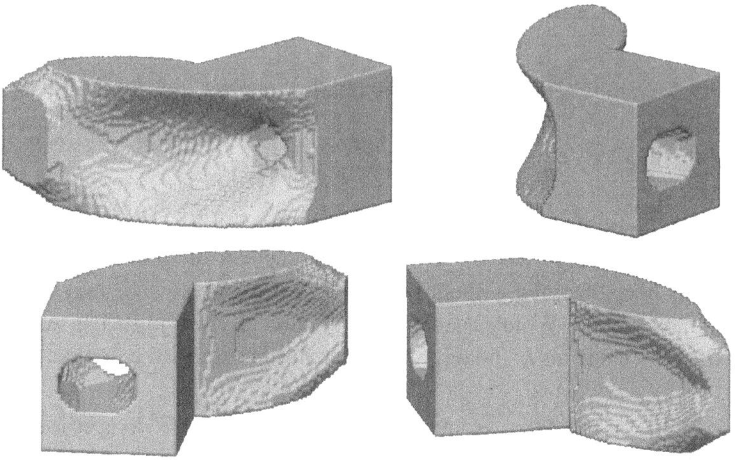

Fig. 9.15 The crank structure of Fig. 9.14 showed from different views. The solution was obtained by Borrvall and Petersson

making this simple problem solvable is to *enlarge* $\mathcal{H}$. This is called *relaxation*. We can replace the old $\mathcal{H}$ by $\mathcal{H} = \{x \in \mathbb{R} \mid x \geq 1\} \cup \{+\infty\}$ and extend the definition of f to this new value, e.g., by letting

$$f(x) = 1/x \quad \text{if } x \in \mathbb{R} \quad \text{and} \quad f(+\infty) = 0.$$

Then the problem of minimizing f for $x \in \mathcal{H}$ has a solution, namely $x = +\infty$. However, the solution has a character which is perhaps not of the sort we are interested in.

Let us return to the archetype problem ($\mathbb{P}_a$). A way of enlarging the set of admissible designs, i.e., doing relaxation, is to allow an infinite perforation by introducing a microscale. This is achieved through *homogenization*. The idea is that every point in the body has a microstructure that is infinitely small and represented by a unit cell. Given such a unit cell there are formulas from which one can calculate global material properties. It turns out that not all types of unit cells result in a proper relaxation in the sense that they give well-posed problems. However, so-called rank 2 microstructure has been shown to give such a problem for compliance minimization. With this microstructure one can obtain a problem for which FE solutions are mesh independent and convergent. Experience shows, however, that resulting solutions consist of large regions of intermediate values for the design density ρ and resembles what is found for the variable thickness sheet problem, see Fig. 9.2. Thus, one is not achieving a topology optimization on the macro scale, which limits the use of this method for such purposes. For an introduction to relaxation and the homogenization approach to topology optimization problems we refer to Bendsøe and Sigmund [4].

9.6 Exercises

Exercise 9.1 Compare the MMA approximation in Sect. 5.3 to the linearization used in deriving the OC method of Sect. 9.1.2. Under what circumstances do the two resulting methods coincide?

Exercise 9.2 Figure 9.16 shows three solutions of the same stiffness optimization problem: an MBB beam where we have taken account of symmetry. One of the solutions has been obtained without restriction, one with a low L^2 bound, and one with a high L^2 bound. Which ones? The L^2 bound is given in (9.10). Among the two obtained by using an L^2 bound, which one has the smallest compliance?

Exercise 9.3 Consider the ground structure in Fig. 9.17. It consists a two-dimensional domain Ω with a thickness $\overline{\rho}$. The load q per unit length is applied on the top surface Γ_1. We want to maximize stiffness, i.e., minimize

$$\int_{\Gamma_1} qu\, dx,$$

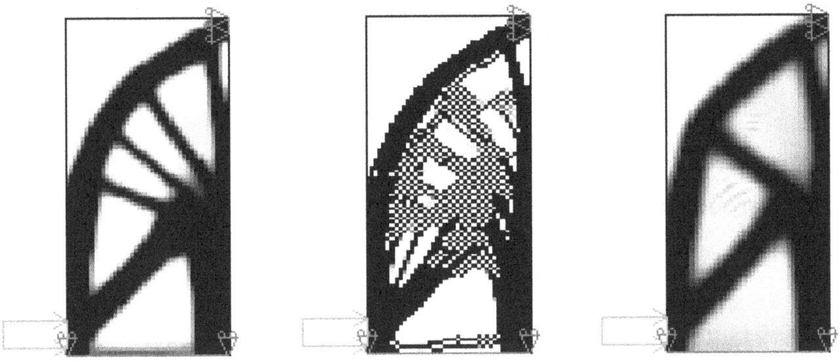

Fig. 9.16 (Color online) The three different optimization results in Exercise 9.2

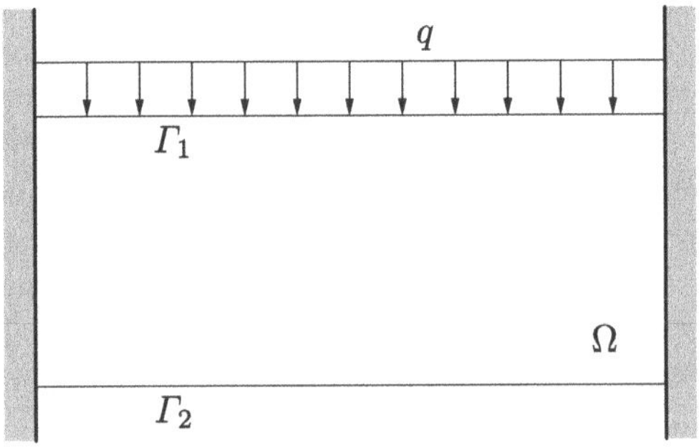

Fig. 9.17 The two-dimensional ground structure of Exercise 9.3

where u is the displacement component in the direction of q, given a constraint on the amount of material that can be used. There are (at least) three different design parameterizations available for doing this optimization:

1. Shape optimization: Find the shape of the lower boundary Γ_2 so as to get an optimal domain $\Omega_{\text{opt}} \subset \Omega$ with thickness $\overline{\rho}$.
2. Thickness optimization: Find an optimal thickness distribution ρ for the domain Ω, where $0 \leq \rho \leq \overline{\rho}$.
3. Topology optimization: Find an optimal thickness ρ of Ω which is such that at each point it is either 0 or $\overline{\rho}$. A proper regularization which gives a well-posed problem is assumed.

Assume that the three problems indicated above can be solved exactly. Which one gives the stiffest structure and which one gives the least stiff structure? Give mathematical arguments for why your answer to the above question is correct.

Exercise 9.4 On the book's homepage (www.mechanics.iei.liu.se/edu_ug/strop/), you may find an extensive computer exercise written by Joakim Petersson where TRINITAS and a Java applet should be used to solve a number of different topology optimization problems.

Answers to Selected Exercises

2.5 a)

$$\alpha^* = 0,\ \beta^* \le \frac{V_0}{Ah}, \quad \text{and} \quad \alpha^* = 1,\ \beta^* \ge 0 \quad \text{arbitrary.}$$

b)

$$\alpha_{\max} = 0.6\colon \alpha^* = 0.2,\ \beta^* = 1.25, \qquad \alpha_{\max} = 0.8\colon \alpha^* = 0.8,\ \beta^* = 2.$$

3.3 a)

$$\begin{cases} \min \dfrac{5}{x_1} + \dfrac{3}{x_2} \\ \text{s.t.} \begin{cases} 25x_1 + \dfrac{3}{5}x_2 - \dfrac{16c_0}{5Pl} \le 0 \\ x_1 \ge 0, \qquad x_2 \ge 0. \end{cases} \end{cases}$$

b)

$$A_1^* = \frac{35}{4}\frac{P^2 l}{Ec_0}, \qquad A_2^* = \frac{7}{4}\frac{P^2 l}{Ec_0}.$$

3.4 a)

$$\begin{cases} \min \dfrac{1}{A_1} + \dfrac{1}{A_2} + \dfrac{2\sqrt{2}}{A_3} \\ \text{s.t.} \begin{cases} A_1 + A_2 + \sqrt{2}A_3 - \dfrac{V_0}{l} \le 0 \\ A_1,\ A_2,\ A_3 \ge 0. \end{cases} \end{cases}$$

b)

$$A_1^* = A_2^* = \frac{1}{4}\frac{V_0}{l}, \qquad A_3^* = \frac{\sqrt{2}}{4}\frac{V_0}{l}.$$

3.5 a)

$$\begin{cases} \min \dfrac{1}{A_1} + \dfrac{2}{A_2} \\ \text{s.t.} \begin{cases} A_1 + A_2 - \dfrac{V_0}{l} \le 0 \\ A_1 \ge \dfrac{V_0}{\alpha l}, \qquad A_2 \ge \dfrac{V_0}{\alpha l}. \end{cases} \end{cases}$$

b)

No solution if $0 < \alpha < 2$,

$$A_1^* = \frac{V_0}{\alpha l}, \qquad A_2^* = \left(1 - \frac{1}{\alpha}\right)\frac{V_0}{l}, \quad \text{if } 2 \le \alpha \le \sqrt{2} + 1,$$

$$A_1^* = (\sqrt{2}-1)\frac{V_0}{l}, \qquad A_2^* = \sqrt{2}(\sqrt{2}-1)\frac{V_0}{l}, \qquad \text{if } \alpha \geq \sqrt{2}+1.$$

3.6 a)

$$\begin{cases} \min \ \dfrac{3}{A_1} + \dfrac{4}{A_2} + \dfrac{5}{A_3} \\ \text{s.t.} \quad \begin{cases} A_1 + A_2 + A_3 - \dfrac{V_0}{l} \leq 0 \\ A_1,\ A_2,\ A_3 \geq 0. \end{cases} \end{cases}$$

b)

$$A_1^* = \frac{\sqrt{3}}{\sqrt{3}+2+\sqrt{5}}\frac{V_0}{l} \approx 0.29\frac{V_0}{l},$$

$$A_2^* = \frac{2}{\sqrt{3}+2+\sqrt{5}}\frac{V_0}{l} \approx 0.34\frac{V_0}{l},$$

$$A_3^* = \frac{\sqrt{5}}{\sqrt{3}+2+\sqrt{5}}\frac{V_0}{l} \approx 0.37\frac{V_0}{l}.$$

3.7 a)

$$\begin{cases} \min \ A_1 + A_2 + \sqrt{2}A_3 \\ \text{s.t.} \quad \begin{cases} \dfrac{3}{A_1} + \dfrac{1}{A_2} + \dfrac{6\sqrt{2}}{A_3} - \dfrac{Eu_0}{Pl} \leq 0 \\ A_1,\ A_2,\ A_3 \geq 0. \end{cases} \end{cases}$$

b)

$$A_1^* = \sqrt{3}(1+3\sqrt{3})\frac{Pl}{Eu_0},$$

$$A_2^* = (1+3\sqrt{3})\frac{Pl}{Eu_0},$$

$$A_3^* = \sqrt{6}(1+3\sqrt{3})\frac{Pl}{Eu_0}.$$

3.8 a)

$$\begin{cases} \min \ A_1 + A_2 + A_3 + A_4 + \sqrt{2}A_5 \\ \text{s.t.} \quad \begin{cases} \dfrac{1}{A_1} + \dfrac{1}{A_2} + \dfrac{4}{A_3} + \dfrac{4}{A_4} + \dfrac{8\sqrt{2}}{A_5} - \dfrac{Ec_0}{P^2 l} \leq 0 \\ A_1, \ldots, A_5 \geq 0. \end{cases} \end{cases}$$

b)

$$A_1^* = A_2^* = \frac{10P^2 l}{Ec_0},$$

$$A_3^* = A_4^* = \frac{20P^2 l}{Ec_0},$$

$$A_5^* = \frac{20\sqrt{2}P^2 l}{Ec_0}.$$

4.4 b)

$$\begin{cases} \min\ (3+2\sqrt{2})\dfrac{1}{x_1} + (3+2\sqrt{2})\dfrac{1}{x_2} + (8+4\sqrt{2})\dfrac{1}{x_3} \\ \text{s.t.} \quad \begin{cases} x_1 + x_2 + \sqrt{2}x_3 - 1 \le 0 \\ x_1,\ x_2,\ x_3 \ge 0. \end{cases} \end{cases}$$

$$x_1^* = x_2^* = \frac{\sqrt{3+2\sqrt{2}}}{2(\sqrt{3+2\sqrt{2}} + \sqrt{2+2\sqrt{2}})} \approx 0.26,$$

$$x_3^* = \frac{\sqrt{1+\sqrt{2}}}{\sqrt{3+2\sqrt{2}} + \sqrt{2+2\sqrt{2}}} \approx 0.34.$$

4.5 b)

$A_1^* = 0.$

c)

$$\begin{cases} \min\ \dfrac{1}{x_2} + \dfrac{1}{x_3} + \dfrac{4}{x_4} \\ \text{s.t.} \quad \begin{cases} x_2 + \sqrt{2}x_3 + x_4 - 1 \le 0 \\ x_1, \ldots, x_4 \ge 0. \end{cases} \end{cases}$$

$$x_2^* = \frac{1}{3\sqrt{\lambda^*}},$$

$$x_3^* = \frac{1}{3\,2^{\frac{1}{4}}\sqrt{\lambda^*}},$$

$$x_3^* = \frac{2}{3\sqrt{\lambda^*}}, \quad \text{where } \sqrt{\lambda^*} = 1 + \frac{2^{\frac{1}{4}}}{3}.$$

6.1 a)
$\partial f/\partial x_j = 2\boldsymbol{u}^T \partial \boldsymbol{u}(\boldsymbol{x})/\partial x_j$. Get $\partial \boldsymbol{u}(\boldsymbol{x})/\partial x_j$ by differentiating $\boldsymbol{K}(\boldsymbol{x})\boldsymbol{u}(\boldsymbol{x}) = \boldsymbol{F}(\boldsymbol{x})$.

6.2
$$\frac{\partial \boldsymbol{K}}{\partial x} = E\begin{bmatrix} A_1\dfrac{2xl_1^2 - 3x^3}{l_1^5} - \dfrac{A_2}{x^2} & -A_1 l\dfrac{l_1^2 - 3x^2}{l_1^5} \\ -A_1 l\dfrac{l_1^2 - 3x^2}{l_1^5} & -\dfrac{3A_1 l^2 x}{l_1^5} \end{bmatrix} \quad \text{etc.}$$

6.3 $x_2 = a + \alpha L$, $y_2 = b$, $l = \sqrt{(a+\alpha L)^2 + b^2}$, $c = \cos\theta = x_2/l$, $\partial c/\partial\alpha = L/l - x_2^2 L/l^3$, $s = \sin\theta = y_2/l = b/l$, $\partial s/\partial\alpha = -bx_2L/l^3$ etc.

6.5
$$\boldsymbol{f}_e^a = \tfrac{1}{6}\rho\omega^2[x_2^2 - 2x_1^2 + x_1x_2 \quad 2x_2^2 - x_1^2 - x_1x_2]^T,$$
$$\partial \boldsymbol{f}_e^a/\partial x_1 = \tfrac{1}{6}\rho\omega^2[-4x_1 + x_2 \quad -2x_1 - x_2]^T,$$
$$\partial \boldsymbol{f}_e^a/\partial x_2 = \tfrac{1}{6}\rho\omega^2[2x_2 + x_1 \quad 4x_2 - x_1]^T.$$

6.7 a)
$$\frac{\partial r}{\partial A_1} = -0.135, \qquad \frac{\partial r}{\partial A_2} = -0.0476.$$

8.7
$$\rho(x) = \frac{6V}{7L}\left[\frac{3}{2} - \frac{x}{L}\left(1 - \frac{x}{2L}\right)\right].$$

8.8
$$\rho(x) = \frac{2V(L-x)}{L^2} \quad \text{for } 0 \le x < L, \quad \text{and}$$
$$\rho(x) = \frac{2V(x-L)}{L^2} \quad \text{for } L \le x \le 2L.$$

8.11
$$\rho(x) = \frac{12V}{h(6M_0L + q_0L^3)}\left[\frac{M_0x}{L} + \frac{q_0x}{2}(L-x)\right].$$

8.12
$$\rho(x) = \frac{V}{2\pi a L}.$$

9.2 The solution in the middle has been obtained without restriction. The left solution has been obtained with a higher value of C_2. This gives a better, i.e. smaller, optimum value than for the right solution.

9.3 Thickness optimization gives the stiffest structure, and shape optimization gives the least stiff structure.

References

1. Achtziger, W.: Topology optimization of discrete structures—an introduction in view of computational and nonsmooth aspects. In: Rozvany, G.I.N. (ed.) Topology Optimization in Structural Mechanics. CISM courses and lectures, vol. 374, pp. 57–100. Springer, Viena (1997)
2. Bazaraa, M.S., Sherali, H.F., Shetty, C.M.: Nonlinear Programming—Theory and Algorithms, 2nd edn. Wiley, New York (1993)
3. Bendsøe, M.P., Klarbring, A.: Joakim Petersson 1968–2002. Struct. Multidiscipl. Optim. **25**, 151–152 (2003)
4. Bendsøe, M.P., Sigmund, O.: Topology Optimization: Theory, Methods and Applications. Springer, Berlin (2003)
5. Bertsekas, D.P.: Nonlinear Programming. Athena Scientific, Belmont (1995)
6. Borrvall, T.: Topology optimization of elastic continua using restriction. Arch. Comput. Methods Eng. **8**, 351–385 (2001)
7. Borrvall, T., Petersson, J.: Topology optimization using regularized intermediate density control. Comput. Methods Appl. Mech. Eng. **190**, 4911–4928 (2001)
8. Borrvall, T., Petersson, J.: Large-scale topology optimization in 3D using parallel computing. Comput. Methods Appl. Mech. Eng. **190**, 6201–6229 (2001)
9. Brockman, R.A.: Geometric sensitivity analysis with isoparametric finite elements. Com. Appl. Numer. Methods **3**, 495–499 (1987)
10. Bruns, T.E., Tortorelli, D.A.: Topology optimization of nonlinear elastic structures and compliant mechnanisms. Comput. Methods Appl. Mech. Eng. **190**, 3443–3459 (2001)
11. Bugeda, G., Oliver, J.: A general methodology for structural shape optimization problems using automatic adaptive remeshing. Int. J. Numer. Methods Eng. **36**, 3161–3185 (1993)
12. Choi, K.K., Kim, N.-H.: Structural Sensitivity Analysis and Optimization 1—Linear Systems. Springer, Berlin (2005)
13. Choi, K.K., Kim, N.-H.: Structural Sensitivity Analysis and Optimization 2—Nonlinear Systems and Applications. Springer, Berlin, (2005)
14. Ehrgott, M., Gandibleux, X. (eds.): Multiple Criteria Optimization: State of the Art Annotated Bibliographic Surveys. Kluwer, Boston (2002)
15. Fleury, C., Braibant, V.: Structural optimization: A new dual method using mixed variables. Int. J. Numer. Methods Eng. **23**, 409–428 (1986)
16. George, P.L.: Automatic mesh generation and finite element computation. In: Ciarlet, P.G., Lions, J.L. (eds.) Handbook of Numerical Analysis, vol. IV. Elsevier, Amsterdam (1996)
17. Gordon, J.G.: Structures or Why Things Don't Fall Down. Penguin, Baltimore (1978)
18. Haftka, R.T., Gürdal, Z.: Elements of Structural Optimization, 3rd revised and expanded edn. Kluwer, Dordrecht (1992)
19. Haslinger, J., Mäkinen, R.A.E.: Introduction to Shape Optimization—Theory, Approximation, and Computation. SIAM, Philadelphia (2003)
20. Hilding, D., Torstenfelt, B., Klarbring, A.: A computational methodology for shape optimization of structures in frictionless contact. Comput. Methods Appl. Mech. Eng. **190**, 4043–4060 (2001)
21. Hughes, T.J.R.: The Finite Element Method—Linear Static and Dynamic Finite Element Analysis. Prentice Hall, Englewood Cliffs (1987) (reprinted by Dover Publications, 2000)
22. Kirsch, U.: Structural Optimization—Fundamentals and Applications. Springer, Berlin (1993)
23. Leung, Y.K., Lo, S.H., Leung, A.Y.T.: Finite Element Implementation. Blackwell Science, Oxford (1996)

P.W. Christensen, A. Klarbring, *An Introduction to Structural Optimization*,

24. Ottosen, N., Petersson, H.: Introduction to the Finite Element Method. Prentice Hall, New York (1992)
25. Petersson, J.: On stiffness maximization of variable thickness sheet with unilateral contact. Q. Appl. Math. **54**, 541–550 (1996)
26. Petersson, J.: A finite element analysis of optimal variable thickness sheets. SIAM J. Numer. Anal. **36**, 1759–1778 (1999)
27. Petersson, J., Sigmund, O.: Slope constrained topology optimization. Internat. J. Numer. Methods Eng. **24**, 359–373 (1998)
28. Piegl, L., Tiller, W.: The NURBS Book, 2nd edn. Springer, Berlin (1997)
29. Rogers, D.F.: An Introduction to NURBS—With Historical Perspective. Academic Press, San Diego (2001)
30. Sigmund, O.: Design of material structures using topology optimization. Ph.D. thesis, DCAMM, Technical University of Denmark (1994)
31. Sigmund, O.: On the design of compliant mechanisms using topology optimization. Mech. Struct. Mach. **25**, 493–524 (1997)
32. Sigmund, O., Petersson, J.: Numerical instabilities in topology optimization: A survey on procedures dealing with checkerboards, mesh-dependencies and local minima. Struct. Optim. **16**, 68–75 (1998)
33. Svanberg, K.: On local and global minima in structural optimization. In: Atrek, E., Gallhager, R.H., Ragsdell, K.M., Zienkiewicz, O.C. (eds.) New Directions in Optimum Structural Design, pp. 327–341. Wiley, New York (1984)
34. Svanberg, K.: The method of moving asymptotes—a new method for structural optimization. Internat. J. Numer. Methods Eng. **24**, 359–373 (1987)
35. Thompson, J.F., Soni, B.K., Weatherill, N.P. (eds.): Handbook of Grid Generation. CRC Press, Boca Raton (1999)
36. Tracy, F.T.: Graphical pre- and post-processor for 2-dimensional finite element programs. Comput. Graph. **11**, 8–12 (1977)

Index